The Attacking Ocean

Also by Brian Fagan

The Attacking Ocean

THE PAST, PRESENT, AND FUTURE OF RISING SEA LEVELS

Brian Fagan

BLOOMSBURY PRESS
NEW YORK • LONDON • NEW DELHI • SYDNEY

Published by Bloomsbury Press, New York

All papers used by Bloomsbury Press are natural, recyclable products made from wood grown in well-managed forests. The manufacturing processes conform to the environmental regulations of the country of origin.

LIBRARY OF CONGRESS CATALOGING-IN-PUBLICATION DATA

Fagan, Brian M.
The attacking ocean : the past, present, and future of
rising sea levels / Brian Fagan.
p.   cm.
Includes bibliographical references and index.
ISBN 978-1-60819-692-0
1. Sea level—History. 2. Ocean—History. I. Title.
GC89.F35 2013
551.45'8—dc23
2012043454

First U.S. Edition 2013

1   3   5   7   9   10   8   6   4   2

Typeset by Westchester Book Group
Printed and bound in the U.S.A. by Thomson-Shore Inc., Dexter, Michigan

To

Atticus Catticus Cattamore Moose

A splendid beast who did everything he could to stop this book being written by dancing on the keyboard at inopportune moments. And he never has to worry about sea levels.

It came to the megacity at dusk, deceptively and unequally . . . In the lowlands, near the seashores, the harbors, the bays, the Sound, the river: apocalypse. The very ocean rose, tsunami-like, relentless, terrifying, bringing devastation by flood and wind and wind-shipped fire, and for some ten million people in a swath a thousand miles wide and encompassing sixteen states, darkness and dread.*

—Hendrik Hertzberg on Hurricane Sandy,
The New Yorker, November 12, 2012

Table of Contents

Alternative Table of Contents

(For those who prefer to explore the narrative geographically)

SOUTH AND SOUTHEAST ASIA, CHINA, AND JAPAN

ISLANDS (ALASKA, PACIFIC, AND INDIAN OCEAN)

NORTH AMERICA

Preface

ALMOST ALL MY LIFE, I've lived by the sea. I've also sailed thousands of kilometers along ocean coasts and across open water, even crossed the Atlantic Ocean. Some of my earliest memories are of lying awake on an English morning, listening to the rain from a southwesterly gale gusting against my bedroom window. Another early childhood experience sticks in my mind—a late 1940s vacation on Jersey, part of the Channel Islands, where the tides run fast with a range of over eleven meters at full or new moon. I remember watching the rising tide course over wide sand flats so fast that you almost had to run to keep ahead of the breakers—not that my father would have allowed me to do anything so rash. He was well aware of the power of a wave born by a powerful tidal stream. In later years, I sailed small yachts in northern European waters, where the direction of tidal streams and the height of the ebb and flood determine which direction you sail and when, and where you'll anchor. Lying aground on a sandbank at thirty-five degrees is no fun, especially if it's the middle of the night and you feel guilty for having misjudged the tide. All these cumulative experiences flooded into my consciousness as I delved into the complex history of rising sea levels over the past fifteen thousand years—since the Ice Age.

You cannot, of course, compare the experience of feeling your way down narrow channels between sandbanks in a small boat with the phenomenon of rising (or falling) sea levels. Tides rise and fall over short cycles of about six hours. In places like Brittany in northern France or the Channel Islands, the landscape changes dramatically from high tide to low. A deep, wide river at high water becomes a narrow stream flowing between rocks and sandbanks at low. It's almost like sailing in a different world, just as it is when you traverse creeks and twisting water-ways in the shallow tidal waters of eastern England, where sandbanks appear and vanish in minutes. The changing sea levels described in these pages are something quite different. We're talking here of gradual, cumulative changes in ocean levels that have risen and fallen for at least 750,000 years and probably much longer. This book is primarily concerned with sea level changes over the past 10,000 years or so and how they have affected humanity.

Think of walking on a sandy beach as the tide is rising. You start at low water as the flood begins. As you walk, you play with the encroaching breakers in bare feet. But, as the hours pass, you find yourself walking farther upslope, often on a much narrower beach. The rise is slow, inexorable, and sure. This is exactly what the gradual sea levels since the end of the Ice Age some fifteen thousand years ago were like, but with one significant difference. There was no ebb. The rise was slow, continual, and cumulative over centuries and millennia, caused by geological processes unfolding thousands of kilometers away.

For the most part, we're unaware of rising sea levels, unless we live along a low-lying coastline, where even a small rise can spread water over a wide area—as happened in the Persian Gulf at the end of the Ice Age, and is the case in places like the Ganges River delta in Bangladesh today. Even in densely populated, threatened areas like the Mekong or the Nile deltas, decadal changes are almost imperceptible. The attacking ocean only enters our urgent consciousness when hurricanes like Katrina, or more recently, Sandy, barrel ashore, bringing high winds, torrential rainfall, and catastrophic sea surges that uproot everything before them. Such sea surges have always assaulted low-lying coasts, but it is only within the past 150 years or so that these vulnerable coasts have become crowded with tens of thousands, even millions, of people.

The past fifteen thousand years have witnessed dramatic sea level changes, which began with rapid global warming at the end of the Ice Age. When the ice started retreating, sea levels were as much as 221 meters below modern levels. Over the next eleven millennia, the oceans climbed in fits and sometimes rapid starts, reaching near-modern levels by about 4000 B.C.E., nine centuries before the first Egyptian pharaohs ruled the Nile valley. This sweeping summary masks a long and complex process of sea level rise, triggered by melting ice sheets, complex earth movements, and myriad local adjustments that are still little understood. By all accounts, these rapid sea level changes had little effect on those humans who experienced them, partly because there were so few people on earth, and also because they were able to adjust readily to new coastlines. Over these eleven thousand years, the world's population was minuscule by today's standards. Fewer than five million people lived on earth fifteen thousand years ago, almost all of them in the Old World. The global population numbered about seven million by six thousand years ago. By today's standards, the world was almost deserted. There was plenty of room to move away from encroaching breakers even in the most densely populated areas. But the marshes and wetlands that protected coasts from storms were still vitally important, not only as natural defenses against the ocean, but also as rich habitats for game, large and small, and also for birds, fish, mollusks, and plant foods.

Global sea levels stabilized about six thousand years ago except for local adjustments that caused often quite significant changes to low-lying places like the Nile delta. The curve of inexorably rising seas flattened out as urban civilizations developed in Egypt, Mesopotamia, and South Asia. Imperceptible changes marked the centuries when Rome ruled the Western world and imperial China reached the height of its power. The Norse explored a North Atlantic identical to ours a thousand years ago; Christopher Columbus and the mariners of the European Age of Discovery sailed over seas that had changed little from the time of the pharaohs. But another variable was now in play, that of the world's population, which climbed rapidly after 4000 B.C.E. By the time of Christ, at least two hundred million people lived on earth. The number had reached a billion or so by the late eighteenth century, when the Industrial Revolution was under way. The densest of these growing populations lived in cities, many of them on river floodplains and low-lying coastal plains. With the rapid expansion of maritime and river-based trade, more and more people settled in strategic coastal locations that became important ports. Now the threat from the ocean increased dramatically, not because of rising sea levels, but from severe weather events like hurricanes and tropical cyclones with their violent sea surges. Tsunamis generated by deep-sea earthquakes assumed much more menacing proportions once people had settled in crowded cities close to the shore. Human vulnerability to a potentially climbing ocean has increased dramatically because there are now so many of us—today seven billion and climbing—at a time when sea level rise has resumed.

The new era of rising sea levels dates from about 1860, the height of the Industrial Revolution. Since then, the world has warmed significantly and the ocean is once again climbing inexorably. Without question, we humans have contributed to the accelerating warming of recent decades. The sea level changes are cumulative and gradual; no one knows when they will end—and the ascent's end is unlikely to come in any of our lifetimes. We live in a very different world from even that of 1860, with tens of millions more people living in coastal cities or farming land only a few meters above sea level. Even a rise of a meter or so will inundate thousands of hectares of rice paddies and major international ports—this before one factors in the savage destruction wrought by sea surges or tsunamis. Our sheer numbers and profound dependence on cargoes transported on the ocean have raised our vulnerability to rising sea levels to a point at which we face agonizing and extremely expensive decisions about flood-control works, sea defenses, or relocation questions that humanity has never wrestled with before.

The Attacking Ocean tells a tale of the increasing complexity of the relationship between humans and the sea at their doorsteps, a complexity created not by the oceans, whose responses to temperature changes and severe storms have changed but little. What has changed is us, and the number of us on earth.

LIKE MY OTHER BOOKS about ancient climate, *The Attacking Ocean* has many narratives and multiple story lines, for the tale covers fifteen millennia of human history, multiple continents, and a great variety of societies—ancient and modern, simple and complex. There is a somewhat chronological gradient in these pages, from earlier societies to later ones. Such an organization is a logical way to explore the tangled history of sea level changes over the past fifteen thousand years, since post–Ice Age warming began. Once the stage is set in chapter 1, which describes how sea levels rise and introduces the dangers of tsunamis and other extreme events, I've divided the story into three parts. "Millennia of Dramatic Change" covers the rapid sea level changes between fifteen thousand years ago and 4000 B.C.E. and their impact on human societies. Chapters 2 to 4 take us from northern Europe to the Black Sea, the Nile valley, and Mesopotamia, through coastal landscapes changed dramatically by fast-climbing sea levels. Chapter 2 brings out a persistent theme—that of the importance of marshes and wetlands to both hunters and subsistence farmers, for these borderlands act as dietary insurance when harvests fail or powerful storms attack low-lying coastlines. In chapter 3, we explore the major environmental changes that resulted from climbing shorelines in the Dardanelles Strait between Europe and Asia. We describe how rising sea levels caused the ponding of the Nile, which led to the creation of the Nile delta, one of the breadbaskets for the Egyptian state. Chapter 4 examines the complex relationship between the early Mesopotamians and a Persian Gulf that turned from a gorge-bisected desert into an arm of the Arabian Sea in a few thousand years, with momentous consequences for humanity. Again, we return to wetlands, this time between the Tigris and Euphrates Rivers, where the god Marduk "laid a reed on the face of the waters" and created one of the first civilizations.

These chapters reside firmly in the past, but the tempo and character of the history changes in "Catastrophic Forces," chapters 5 to 10. Now we are much closer to the present, telling stories of human vulnerability and of societies that often have direct relationships with those of today. Here, most chapters move seamlessly from the past into the present, for the same processes that shaped relationships with ancient sea levels still operate today. For instance, countless early Mediterranean ports had problems with natural sinking and also with accumulating river silt and severe storms. The same problems afflicted medieval Venice and continue to threaten its future today—which is why we journey from past into the present in this and other chapters. The same argument applies to chapters 9 and 10, which describe sea level changes in China and

Japan. In all these chapters, the emphasis is more on major natural events, which bring catastrophe in their train, especially sea surges and tsunamis, the latter the dominant theme of chapter 10. In the case of northern Europe and south Asia, I bring the story forward to later times, leaving the modern struggles with rising sea levels to later chapters.

"Challenging Inundations" comprises the final five chapters, in which I describe some of the extraordinary challenges that confront us today. Again, the coverage is geographic, largely because each area has its own distinctive issues. Chapter 11 describes the menacing difficulties that confront Bangladesh, where 168 million people live close to sea level with nowhere to go. Here, the long-term problem of environmental refugees raises its head, a global problem that is currently on few governments' agendas, although it should be. Chapter 12, "The Dilemma of Islands" raises not only the issue of environmental refugees, but also the difficulties of relocating entire villages and small island nations. "The Crookedest River in the World" takes us back deep into the history of the Mississippi and its peoples to a delta coast and modern cities threatened from both up- and downstream. This is a story of sea defenses erected in the face of extreme storms, unpredictable river floods, and rising sea levels. The Low Countries face the same dilemma as Bangladesh, lands walled off in an ultimate expression of human determination to resist the attacking sea to the death. The epilogue looks at the future and rising sea levels in the context of the United States, where millions of people live in houses but a few meters above today's sea level. Hurricane Sandy, which devastated much of New York and the New Jersey Shore in October 2012, provided a sobering reminder of just how vulnerable we are to a future of more frequent extreme weather events and violent sea surges that can render tens of thousands homeless.

Everyone should start this book by reading chapter 1, but thereafter you have choices. Brief summary statements introduce each section to help you on your way. You can follow the story chapter by chapter, moving from area to area as the narrative moves forward. If you find this confusing, especially the winding back of chronological narratives from one chapter to the next, you can explore different areas by jumping between widely dispersed chapters. An alternative table of contents on page xi guides you through this option. This is a particularly effective way of moving from past to present within specific geographies. Provided you end by reading chapter 15, you should leave both with focused narratives of areas that interest you and with a general sense of the central messages of the book. If you choose this approach, your choices are open-ended. Using the notes at the end of the book, you can then delve more deeply into the enormous literature that surrounds each chapter.

Author's Note

GEOGRAPHICAL PLACE NAMES ARE spelled according to the most common usage. Archaeological and historical sites are presented as they appear most commonly in the sources I used to write this book. Some obscure locations are omitted from the maps for clarity. Interested readers should consult the specialist literature.

The notes tend to emphasize sources with extensive bibliographies to allow you to enter the more specialized literature if desired.

The B.C.E./C.E. convention is used throughout this book. The "present" by international agreement is 1950 C.E. Following routine practice, dates before twelve thousand years ago appear as years before present.

After considerable debate, we (my editors and I) decided that we would use metric measurements in these pages, to simplify the narrative. This is because most science now employs metric conventions. For those who are bewildered, a mile is 1.6 kilometers, a foot is 0.3 meter, and an inch is 2.54 centimeters. An acre is 0.4 hectare. Doubtless I will hear in short order from those who refuse to think of their world in metric terms. It will only take you a few seconds with your calculator or computer to convert any measurement, however esoteric.

A knot (a nautical mile), commonly used on charts and in sailing directions, is 1.85 kilometers. I use it here to refer to boat speeds and to the velocity of currents and tides, the usual nautical practice.

All radiocarbon dates have been calibrated to dates in calendar years using the latest version of what is a constantly revised calibration curve. You can view the calibration curve at www.calpal.de.

Following common maritime convention, wind directions are described by the direction they are blowing *from*. For example, a westerly wind blows from the west, and northeast trade winds from the northeast. Ocean currents and tides, however, are described by the direction they are flowing *toward*. Thus, a northerly wind and a northerly tide flow in opposite directions.

Minus One Hundred Twenty-Two Meters and Climbing

ON OCTOBER 28, 2012, Hurricane Sandy, the largest Atlantic hurricane on record, came ashore in New Jersey. Sandy's assault and sea surge brought the ocean into neighborhoods and houses, inundated parking lots and tunnels, turned parks into lakes. When it was all over and the water receded, a huge swath of the Northeast American coast looked like a battered moonscape. Only Hurricane Katrina, which devastated New Orleans in 2005, was more costly. Katrina, with its gigantic sea surge, had been a wake-up call for people living on low-lying coasts, but the disaster soon receded from the public consciousness. Sandy struck in the heart of the densely populated Northeastern Corridor of the United States seven years later and impacted the lives of millions of people. The storm was an epochal demonstration of the power of an attacking ocean to destroy and kill in a world where tens of millions of people live on coastlines close to sea level. This time, people really sat up and took notice in the face of an extreme weather event of a type likely to be more commonplace in a warmer future. As this book goes to press, a serious debate about rising sea levels and the hazards they pose for humanity may have finally begun—but perhaps not.

Sandy developed out of a tropical depression south of Kingston, Jamaica, on October 22. Two days later, it passed over Jamaica, then over Cuba and Haiti, killing seventy-one people, before traversing the Bahamas. Come October 28, Sandy strengthened again, eventually making landfall about 8 kilometers southwest of Atlantic City, New Jersey, with winds of 150 kilometers an hour. By then, Sandy was not only an unusually large hurricane but also a hybrid storm. A strong Arctic air pattern to the north forced Sandy to take a sharp left into the heavy populated Northeast when normally it would have veered into the open Atlantic and dissipated there. The blend produced a super storm with a wind diameter of 1,850 kilometers, said to be the largest since 1888, when far fewer people lived along the coast and in New York. Unfortunately, the tempest also arrived at a full moon with its astronomical high tides. Sandy was only a Category 1 hurricane, but it triggered a major natural disaster partly because it descended on a densely populated seaboard where thousands of houses and other property lie within a few meters of sea level. Imagine the destruction a Category 5 storm would have wrought—something that could happen in the future.

The scale of destruction was mind-boggling. Sandy brought torrential downpours, heavy snowfall, and exceptionally high winds to an area of the eastern United States larger than Europe. Over one hundred people died in the affected states, forty of them in New York City. The storm cut off electricity for days for over 4.8 million customers in 15 states and the District of Columbia, 1,514,147 of them in New York alone. Most destructive of all, a powerful, record-breaking 4.26-meter sea surge swept into New York Harbor on the evening of October 29. The rising waters inundated streets, tunnels, and subways in Lower Manhattan, Staten Island, and elsewhere. Fires caused by electrical explosions and downed power wires destroyed homes and businesses, over one hundred residences in the Breezy Point area of Queens alone. Even the Ground Zero construction site was flooded. Fortunately, the authorities had advance warning. In advance of the storm, all public transit systems were shut down, ferry services were suspended, and airports closed until it was safe to fly. All major bridges and tunnels into the city were closed. The New York Stock Exchange shut down for two days. Initial recovery was slow, with shortages of gasoline causing long lines. Rapid transit systems slowly restored service, but the damage caused by the storm surge in lower Manhattan delayed reopening of critical links for days.

The New Jersey Shore, an iconic vacation area in the Northeast, suffered worst of all. For almost 150 years, people from hot, crowded cities have flocked to the Shore to lie on its beaches, families often going to the same place for generations. They eat ice cream and pizza, play in arcades once used by their grandparents, drink in bars, and go to church. The Shore could be a seedy place, fraught with racial tensions, and sometimes crime and violence, but there was always something for everybody, be they a wealthy resident of a mansion, a contestant in a Miss America pageant, a reality TV actor, a skinny-dipper, or a musician. Bruce Springsteen grew up along the Shore and his second album featured the song "4th of July, Asbury Park (Sandy)," an ode to a girl of that name and the Shore. "Sandy, the aurora is rising behind us; the pier lights our carnival life forever," he sang. The words have taken on new meaning since the hurricane came.

Fortunately, the residents were warned in advance of the storm. They were advised to evacuate their homes as early as October 26. Two days later, the order became mandatory. New Jersey governor Chris Christie also ordered the closure of Atlantic City's casinos, a decision that proved wise when Sandy swept ashore with brutal force, pulverizing long-established businesses, boardwalks, and homes. Atlantic City started a trend when it built its first boardwalk in 1870 to stop visitors from tracking sand into hotels. Boardwalk amusements are big business today, many of them faced by boardwalks that are as much as a 0.8-kilometer from the waves. Now many of the Shore's iconic boardwalks are history. The waves and storm surge destroyed a roller coaster in Seaside Heights; it lay half submerged in the breakers. Seaside Heights itself was evacuated because of gas leaks and other dangers. Piers and carousels vanished; bars and restaurants were reduced to rubble. Bridges to barrier islands buckled, leaving residents unable to return home. The Shore may be rebuilt, but it will never be the same. A long-lived tradition has been interrupted, perhaps never to return. For all the fervent vows that the Shore will rise again, no one knows what will come back in its place along a coastline where the ocean, not humanity, is master.

As the waters of destruction receded, they left $50 billion of damage behind them, and a sobering reminder of the hazards millions of people face along the densely populated eastern coast of the United States.

Like Hurricanes Katrina in 2005 and Irene in 2011, Sandy showed us in no uncertain terms that a higher incidence of extreme weather events with their attendant sea surges threaten low-lying communities along much of the East Coast—from Rhode Island and Delaware to the Chesapeake and parts of Washington, DC, and far south along the Carolina coasts and into Florida, which escaped the full brunt of Sandy's fury. There, high winds and waves washed sand onto coastal roads and there was some coastal flooding, a warning of what would certainly occur should a major hurricane come ashore in Central or Southern Florida—and the question is not *if* such an event will occur, but *when*.

ONE HUNDRED AND twenty meters and climbing: that's the amount of sea level rise since the end of the Ice Age some fifteen thousand years ago. Slowly, inexorably, the ascent continues in a warming world. Today the ocean laps at millions of people's doorsteps—crouched, ready to wreak catastrophic destruction with storm-generated sea surges and floods. We face a future that we are not prepared to handle, and it's questionable just how much most of us think about it. This makes the lessons of Katrina, Irene, and Sandy, and other recent storms important to heed. Part of our understanding of the threat must come from an appreciation of the complex relationship between humanity and the rising ocean, which is why this book begins on a low land bridge between Siberia and Alaska fifteen thousand years ago . . .

BETWEEN SIBERIA AND Alaska, late summer, fifteen thousand years ago. A pitiless north wind fills the air with fine dust that masks the pale-blue sky. Patches of snow lie in the shallow river valleys that dissect the featureless landscape. A tiny group of humans trudge down the valley close to water's edge, the wind at their backs, the men's eyes constantly on the move, searching for predators. They can hear the roar of the ocean in the shallow bay, where wind squalls whip waves into a white frenzy. A few days earlier, the women had trapped some arctic ptarmigan with willow snares, but the few remaining birds hanging at their belts are barely enough for another meal. A dark shadow looms through the dusty haze—a solitary young mammoth struggling to free itself from mud at river's edge.

The men fan out and approach from downwind, scoping out the prospects for a kill. The young beast is weakening rapidly after days in the muddy swamp. Nothing is to be gained by going in for the kill at the moment, so the band pitches camp a short distance away and lights a large fire to keep away predators. A gray, bitterly cold dawn reveals the helpless mammoth barely clinging to life, mired up to its stomach. A young man leaps onto the beast's hairy back and drives his stone-tipped spear between its shoulder blades, deep into the heart. He jumps off to one side, landing in the mud. The hunters watch the mammoth's death throes and thrust more spears into their helpless prey. Soon everyone moves in to skin the flanks and dismember the exposed parts. A short distance away, wolves lurk, ready to move in when the humans leave.

Back in camp, the men build low racks of fresh mammoth bone and lay out strips of flesh to dry in the ceaseless wind, while the women and children cook meat over the fire. Around them, the dust-filled gloom never lifts, the wind blows, and the roar of the ocean never leaves their consciousness. The sea is never a threat, for their lives revolve around the land and they can easily avoid any encroaching waves by doing what they always have done—keeping on the move.

THIS IMAGINED MAMMOTH hunt unfolded at a time when the world was emerging from a prolonged deep freeze. The bitter cold of a long glacial cycle had peaked about seven thousand years earlier, the most recent of a more than 750,000-year seesaw of lengthy cold and shorter inter-glacial periods driven by changes in the earth's orbit around the sun, which had began 2.5 million years ago.[1] Twenty-one thousand years ago, world sea levels were just under 122 meters below modern shorelines. The seas were beginning to rise fast, as a rapid thaw began and glacial meltwater flowed into northern oceans. Soon one would need a skin boat to cross from Siberia to Alaska and the mammoth hunters' killing grounds would be no more.

An ascent of 122 meters is a long way for oceans to climb, but climb it they did, most of it with breathtaking rapidity by geological standards, between about fifteen thousand years ago and 6000 B.C.E. Most of the ascent resulted from powerful meltwater pulses that emptied enormous quantities of freshwater from ice sheets on land into northern waters and around Antarctica. This was not, of course, the first time that such a dramatic rise had transformed an ice-bound world, but there was an important difference fifteen millennia ago. For the first time, significant numbers of human beings, perhaps as many as hundreds of thousands of them, lived in close proximity to the ocean.

Some traveled offshore. Fifty thousand years ago, even while the late Ice Age was at its height, small numbers of Southeast Asians had already ventured into open tropical waters to what are now Australia and New Guinea. Well before twenty thousand years ago, people were living on the islands of the Bismarck Strait in the southwestern Pacific.[2] These voyages took place long before melting ice sheets and rising sea levels transformed the Ice Age world of *Homo sapiens*.

WE LIVE IN a rapidly warming world, where human activities now play a significant part in long-term climate change and have done so since the Industrial Revolution, when fossil fuels like coal came into widespread use. It's hard for us to imagine just how different the world was twenty-one thousand years ago. Much of it lay under thick ice. Two huge ice sheets covered virtually all of North America, from the Atlantic to the Pacific. The Cordilleran ice sheet, centered on the Rockies and western coastal ranges, mantled 2.5 million square kilometers. The enormous Laurentide ice sheet lapped the Cordilleran in the west and covered over 13 million square kilometers of what is now Canada. It was nearly 3,353 meters thick over Hudson Bay. Its southern extremities covered the Great Lakes and penetrated deep into today's United States. The Greenland ice sheet was 30 percent larger than today. Another smaller ice sheet linked it to the northern margins of the Laurentide.

In northern Europe, the Scandinavian ice sheet extended from Norway to the Ural Mountains over an area of 6.6 million square kilome-

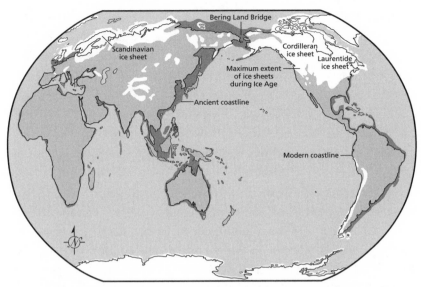

Figure 1.1 *Map showing approximate extent of ice sheets and lower sea levels during the late Ice Age.*

ters, may even have reached Spitsbergen, and flowed over much of the north German Plain. A smaller ice sheet covered about 340,000 square kilometers and reached halfway down the British Isles. Glaciers descended close to sea level in the Southern Alps. In Siberia and Northeast Asia, ice extended over at least ten times the area of the British ice sheet. Extensive ice sheets mantled the Himalayas.

The Antarctic ice sheet was about 10 percent larger; seasonal sea ice extended eight hundred kilometers out from the continent. There were important ice sheets on the Andes Mountains, in South Africa, southern Australia, and New Zealand. Twenty-one thousand years ago, there was two and a half times as much ice on land as there is today. Of that, 35 percent was on North America, 32 percent on Antarctica, and 5 percent on Greenland. Today, 86 percent of the world's continental ice is on Antarctica, 11.5 percent on Greenland.

There was so much water locked up in glacial ice sheets and sucked out of the oceans that global sea levels were up to 122 meters below those of today. These much lower sea levels changed the shape of entire continents. Perhaps most significant historically was the bitterly cold and low-lying Bering Land Bridge that linked Siberia and Alaska, a natural highway that brought the first humans to the Americas. Dry land joined islands in Southeast Alaska and the Pacific Northwest of North America. Much farther south, San Francisco's Golden Gate channel was a narrow tidal gorge with fast-moving rapids. Continental shelves extended some distance off the Southern California coast, leaving but eleven kilometers of open water between the mainland and the Channel Islands close offshore.

On the other side of the Pacific, low sea levels joined the Japanese islands to Sakhalin Island in the north and brought them much closer to the Chinese and Korean mainland. In China, major rivers like the Huang He in the north and the Yangtze in the south flowed through incised, narrow valleys rather than broad floodplains. Rolling plains stretched far into the distance off Southeast Asia. Only short stretches of open water separated the mainland from Australia and New Guinea, which were a single landmass, now covered by the shallow Arafura Sea.

The configuration of the Indian Ocean was much different from today. Bangladesh lay far above sea level by modern standards, incised by the Ganges and other rivers that flowed much more rapidly to the sea. Sri Lanka's twenty-nine-kilometer-long Rama's Bridge, now a chain of limestone shoals, was a land bridge to India. The Persian Gulf was dry land, an arid landscape bisected by a narrow gorge that drained the highlands and plains at its head.

Had one looked down from a satellite at Europe and the Mediterranean eighteen thousand years ago, one would have surveyed unfamiliar landscapes. Continental shelves extended far into the Bay of Biscay. You could walk from Britain to France, had you possessed a canoe to carry you across an enormous estuary that carried the waters of the Rhine, Seine, and Thames Rivers of today. The southern North Sea was a land of shallow lakes and marshes. The Mediterranean was far smaller, its narrow entrance at the Strait of Gibraltar scoured by fast-running currents. The northern Aegean Sea ended in a high barrier that isolated what is now the Black Sea from the ocean. The Euxine Lake, formed by glacial and freshwater runoff from the north, lay behind the natural berm. On the other side of the Mediterranean, the arid Nile delta with its sand dunes extended far into what is now open sea.

Everywhere large rivers like the Thames and the Rhine had lower courses and estuaries far different from those of today. The Nile flowed through a twisting, narrow gorge, where the annual flood for the most part remained close to the river channel rather than spilling over a wide floodplain as it did until the building of the Aswan Dam. In the Americas, the St. Lawrence River did not exist; it was under the Laurentide ice sheet. The Mississippi and Amazon Rivers cut far below their modern gradients, with almost none of the ponding and wetland formation that developed as sea levels rose.

Rapid, natural global warming transformed the late Ice Age world into what was effectively an entirely different place in less than ten thousand years. Within this brief time frame, the world's sea levels rose 122 meters.

EUSTACY AND ISOSTASY: the words used to describe sea level changes glide easily off the tongue, but they mask very complex and still only partially understood geological processes. What does cause the world's sea levels to rise and fall? Isostatic changes result from local upward and downward shifts in the lithosphere, the uppermost layers of the earth. Such factors as earthquake activity and shifts of tectonic plates far below the earth's surface are important contributors to sea level change. Subsidence in river deltas, changes in glaciers, even sediment compaction—anything that adds to or subtracts from the weight of the earth's crust—all can cause isostatic sea level rises, such as are common in places like Shanghai.[3]

Eustatic, global sea level rise is completely different, a measure of the increase in the volume of water in the oceans expressed as a change in water height. Everyone knows that water expands as it heats. When the earth's atmosphere warms, the ocean absorbs much of the increasing heat and its waters swell. Thermal expansion is the major cause of global sea level rise since the 1860s, when the Industrial Revolution with its promiscuous use of fossil fuels added more carbon and other pollutants to the atmosphere—in other words, when humanly caused global warming began. At present, eustatic sea level rise advances at a rate of about two millimeters a year if calculated on an average of the past century. Over the past fifteen years, however, the averaged rate is around three millimeters a year, apparently a direct, accelerated response to global warming.

In recent years, we've learned a great deal about the world's glaciers, thanks to satellite technology. We can now measure the elevation of glaciers, their growing and shrinking masses from space with satellite technology, a task hitherto accomplished by arduous fieldwork on the ground. We can also measure the velocity of moving ice and establish the grounding points of glaciers. All of this gives us a more complete overview of the world's ice. The new portraits show that the Antarctic and Greenland ice sheets have contributed only a tiny proportion to annual sea level rise—until recently. However, ice sheets are now contrib-

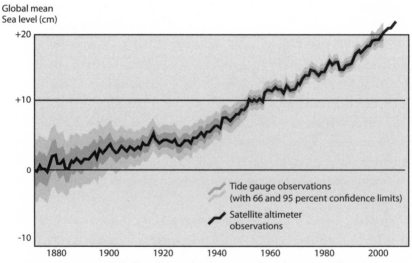

Figure 1.2 *A global average of tide gauge data from 1870 to 2000. The new satellite altimeter data is superimposed from 1990. Data from both sources is very similar, chronicling a more rapid sea level rise since 1990. Courtesy: NASA.*

uting double the amount of water they did in the recent past. All the signs are that the rate is accelerating, perhaps dramatically.[4]

Glaciers retreat most rapidly when they end in water. The Mendenhall Glacier near Juneau has retreated five kilometers since 1760 and a kilometer since 2000. Muir Glacier in Glacier Bay has receded ten kilometers since 1941, the Columbia Glacier in Prince William Sound by thirteen since 1981. The general retreat contrasts dramatically with that of the Little Ice Age (ca. 1350 to 1850 C.E.), when Alaska's mountain glaciers not only thickened but also advanced.

The Greenland ice sheet covers 4,550 million square kilometers. If this were to melt completely, global sea levels would rise by seven meters. Recent estimates using a variety of high-technology measurements and satellite data place the annual loss of Greenland ice at about 96 cubic kilometers in 1996, and almost three times that, at 290 cubic kilometers a year, in 2006. Most of this acceleration is occurring in the southernmost glaciers, but the melting is gradually moving northward. Were the entire Greenland ice sheet to melt, which has happened in the remote past, what is now a frozen continent would become an archipelago of islands surrounding a central sea—until crustal adjustments caused the continent to rebound from the weight imposed on it by the ice.

Antarctica at the other end of the world is a vast desert continent, 98 percent of which is covered with ice. Until about 2000, the amount of Antarctic ice was slowly increasing. But recent satellite measurements show that the ice sheets are now losing mass. The East Antarctic ice sheet lies above sea level and is relatively stable. If it were to melt, the resulting sea level rise would be in the order of fifty meters. Much of the West Antarctic ice sheet is aground on the seabed, where changing ocean temperatures affect the stability of its foundations. When these decay, the upper parts of the ice sheet can detach and move seaward. Such melting occurred at the end of the last Ice Age and may well occur again in the near future. Should the entire West Antarctic ice sheet melt, world sea levels would rise by about five meters. If the sheet were to vanish within twelve hundred years, the ocean would climb about thirty to fifty centimeters per century. A five-hundred-year meltdown would cause sea level rises of as much as one hundred centimeters a century.

Predicting sea level changes is a game of geological poker. Unfortunately it is nearly impossible to establish whether changes in ice sheets are mere decadal events, part of normal fluctuations, or something that portends much more dramatic changes in the future. Many experts believe that a two-meter rise by 2100 should be the basis for future planning. Geoscientists Orrin Pilkey and Rob Young believe that "coastal management and planning should be carried out assuming that the ice sheet disintegration will continue and accelerate. This is a cautious and conservative approach."[5] Pilkey and Young are sober voices in what often becomes a morass of alarmism and sensationalist headlines. Unfortunately, impending (if probably fictional) climatological catastrophe makes for headline news.

WHAT, THEN, DO we know about sea level rise since the end of the Ice Age? So many local factors are involved that it's hard to provide anything more than a general portrait of global sea level changes over the past fifteen thousand years—and even then the experts disagree. Here's a generalized framework, based on many data sources, which will provide a reference point for later chapters.[6]

When the great thaw began, ice sheets retreated precipitously by geological standards. Enormous volumes of meltwater cascaded into northern waters, much of it in dramatic accelerations in sea level rise (often called "pulses") recorded in Caribbean corals and other sources. There were at least four major pulses, the first about 19,000 years ago, when the sea rose between ten and fifteen meters within a mere five centuries. Another major meltwater release, most likely from North American ice sheets, came between 14,600 and 13,600 years ago. This time the sea level rose between sixteen and twenty-four meters. These major sea level rises came before the onset of a 1,300-year cold snap, the so-called Younger Dryas of 10,800 to 9500 B.C.E. (named by geologists after a polar flower), when sea level rise slowed, only to resume sharply with another meltwater pulse between 9500 and 9000 B.C.E. A fourth meltwater pulse between 6200 and 5600 B.C.E. brought a minor rise, perhaps a meter or so.

By 4000 to 3000 B.C.E., global sea level rise had virtually ceased, except for local crustal adjustments as a result of melting ice sheets. Many islands and coastal areas far from glaciated areas experienced sea levels several meters higher than they are today. The earth's crust responded locally to changes in ice coverage and water loading by siphoning water away from equatorial ocean basins into depressed areas close to vanished ice sheets. At the same time, the weight of increased amounts of meltwater affected continental shelves by tilting the shoreline upward and lowering local sea levels. Despite these changes, the rate of global sea level rise remained very low until the mid-nineteenth century C.E.

Now the situation has changed significantly. Data from coastal sediments, tidal gauges, and satellites tell us that sea levels have been rising once more since the late nineteenth and early twentieth centuries. Readings from the altimeter aboard the TOPEX/Poseidon satellite record an even higher rate of about 2.8 millimeters annually in recent years, hinting at a long-term acceleration of sea level rise.[7] Satellites also tell us that the Greenland and Antarctic ice sheets are discharging ice into the oceans more rapidly. But with the kinds of temperature-rise projections of 2 to 5 degrees Celsius projected for the twenty-first century, the resulting virtually total meltdown would take several hundred years. Nevertheless, even with possible accelerated discharge from the West Antarctic ice sheet, it seems unlikely that the rate of sea level rise would exceed those of major post–Ice Age meltwater pulses.

Comforting words, perhaps, but the sea level rises of thousands of years ago came at a time when global populations were a fraction of those of today. Even as recently as five thousand years ago, the most concentrated urban populations in nascent cities like Uruk in Mesopotamia would have numbered less than ten thousand people. Elsewhere, in a world still peopled by small farming communities and hunting bands that were normally on the move, population densities never would have numbered more than a few people per square mile. The world was far from congested, so adjusting to rapid sea level rise was either a matter of shifting camp or of clearing new land for farming villages that were, in any case, often rebuilt or moved every few generations. Today, with cities in the millions and billions of coastal dwellers living at sea level or close to it, the long-term challenges of accelerated sea level rise are already making themselves felt in a much more vulnerable world. As many experts have testified, increased warming brings a higher incidence of extreme hurricanes and severe gales, and also of tropical cyclones and their sea surges, quite apart from the terrible consequences of tsunamis triggered by earth movements on the ocean floor that devastate coastal settlements. These may be short-term events, but they are deadly. Witness the notorious Lisbon earthquake and tsunami of 1755.

Lisbon, Portugal, November 1, 1755. All Saints' Day was the most important religious holiday in the calendar. Everyone—rich and poor, young and old—flocked to cathedrals and churches, for not to attend was to open oneself to accusations of heresy, no light matter in eighteenth-century Portugal.[8] People jostled for space; overflow crowds spilled into the streets. Most of the foreigners in the cosmopolitan city stayed at home. As the hymns and prayers were at their height, a loud rumbling and trembling from deep beneath the earth drowned out the sounds of worship, for all the world like distant peals of thunder or the wheels of a passing heavy carriage. Then came the shaking, so violent that houses collapsed in seconds. Towers and spires swayed uncontrollably; church bells pealed in terrible cacophony, then tumbled to the ground. Entire congregations met their deaths under collapsing roofs and church walls. Huge chunks of masonry and jagged rubble filled alleyways and streets, burying helpless, screaming victims. Dense clouds of dust obscured the blue sky of what had been a lovely sunny day. Within minutes, Lisbon was a rubble field, ravaged by fires started by church candles and scattered hearths.

The survivors fled to open ground, away from tottering buildings and raging fires. Many headed for the open banks of the Tagus River, where they stood in shock, convinced that the Day of Judgment had come. They loudly begged for divine mercy. Priests moved through the crowds urging them to repent for their sins in the face of God's wrath. How else could such a catastrophe have descended on them on the holiest of days? There were, at the time, no plausible scientific explanations.

Figure 1.3 *The Lisbon earthquake and tsunami as depicted in Georg Ludwig Hartwig's* Volcanoes and Earthquakes: A Popular Description, *published in 1887. Author's collection.*

Ninety minutes after the earthquake, the crowds at water's edge saw the Tagus rock and roll. Ships at anchor offshore gyrated wildly. Then "there appeared at some distance a large body of water, rising as it were like a mountain; it came foaming and roaring, and rushed towards the shore with such impetuosity, that we all immediately ran for our lives."[9] The roaring came from a huge tsunami wave, said to have been at least twelve meters high, which swept up the river and surged ashore as spectators fled. The three thousand or so people standing on a new stone wharf in hopes of finding a boat drowned when the quay tipped over. The wall of water rushed as far as 2.4 kilometers inland, inundating buildings, overthrowing bridges, and dashing thousands of helpless onlookers against buildings and walls. Then the wave receded precipitously, carrying hundreds more victims to their deaths and uncovering areas of the riverbed that were normally twelve meters underwater. Oceangoing ships lay helplessly aground. The surviving onlookers moved to the bank, gasping at the sight of fish flapping helplessly. Ten minutes later a second even more violent fifteen-meter wave advanced up the estuary and also cascaded ashore, promptly followed by a third, both traveling "like a torrent, tho' against the wind and tide."[10] The waves moved so fast that galloping horses could barely escape them.

As many as twenty thousand to sixty thousand people died in the Lisbon earthquake and tsunami. Entire cities in southern Portugal and Spain fell victim to the waves. Tsunami waves swept ashore as far away as Morocco and the island of Madeira (520 kilometers) west of Africa in the Atlantic. Nine hours later, 3.6-meter breakers flooded lowlands on Caribbean islands. Northern Europe felt the effects of the tsunami, the first global natural disaster that prompted the first serious research into such events.

Tsunamis are unpredictable events generated by invisible earth movements, such as earthquakes caused by the collision of tectonic plates far below the sea surface. The word "tsunami" itself is a Japanese term meaning "harbor wave," which dates back at least four centuries and is said to have been coined by fishermen when they returned to devastated harbors after not even having noticed tsunami waves in deep water. Great tsunamis result from extensive displacements of the seafloor, perhaps over hundreds of kilometers, over distances longer than the depth of the water. The waves generated by these enormous earthquakes are extremely long and travel great distances at speeds up to 640 kilometers an hour.[11] They are far more damaging than smaller scale tsunamis caused by more local events such as a underwater landslide. In such cases, big waves may result, but they soon dissipate. Large tsunami waves are immensely powerful and quite different from the conventional breakers beloved by surfers. They are solid walls of water that sweep everything before them and rush ashore until friction or gravity cause them to slow and recede.

The Lisbon tsunami of 1755 was not unprecedented. We know, for example, that at least eight large tsunamis have struck the coasts of Spain, Portugal, and Morocco over the past twelve thousand years, at intervals of about fifteen hundred years. In 6100 B.C.E., the Storegga underwater landslide displaced vast amounts of seawater off western Norway and caused a tsunami as far away as the Orkney Islands off northern Scotland.[12] Then there is what one might call the mother of all historical natural disasters, the great eruption and resulting tsunami that blew much of Santorini Island in the Aegean into space in about 1627 B.C.E. An entire town, now known as Akrotiri, vanished under a cloud of ash and pumice.[13] The inhabitants must have had some warning, for no skeletons lie among its ash-smothered dwellings, which stand up to three stories high. Wine jars, storage pots, the remains of a bed, and bright friezes are all that remains of a once-vibrant community. On the walls, a fisherman returns home with his catch; two boys exchange fisticuffs. Fast ships with serried oarsmen pass by a town amid a pod of dolphins.

A visit to Akrotiri is a stroll through a moment frozen in time. One can imagine the inhabitants grabbing their possessions, driving bleating goats into boats, and rowing hastily away as lumps of pumice drop into the seething water. Then a sudden explosion and oblivion, and a once-prosperous town was forgotten until Greek archaeologist Spyridon Marinatos unearthed some of its houses and alleyways in 1967. The scale of the explosion boggles the mind. What had once been one island measuring about nine by six kilometers became four small ones. Ash from the eruption fell over a large area, some of it on Crete, 177 kilometers to the south, at the time the center of Minoan civilization with its far-flung trade networks, extensive olive groves, and wealthy palaces. A tsunami after the eruption lashed the Cretan shoreline with huge waves, which must have caused considerable damage and disrupted mercantile activity over a vast area. Many experts believe the surging ocean permanently weakened Minoan civilization.

When visiting the deep Santorini crater, one's mind turns to Plato's account of the lost continent of Atlantis immortalized by the Greek philosopher with his tale of kings "of great and marvelous power," overthrown by "portentous earthquakes and floods."[14] Despite enduring searches by the obsessed, Atlantis is almost certainly a figment of classical imagination and never existed, perhaps a folk memory of the Santorini cataclysm or some other tsunami. The Greek historian Thucydides witnessed an earthquake at Orobiae in the Euboian Gulf off eastern Greece in 429 B.C.E. He recorded how the sea, "retiring from the then line of coast, returned in a huge wave and invaded a great part of the town, and retreated leaving some of it still under water, so what was once land is now sea; such of the inhabitants perishing as could not run up to the higher ground in time."[15]

Really major tsunamis like the Lisbon event have global consequences. The earthquake and tsunami that hit Shimoda, south of Tokyo in Japan on December 1, 1854, brought small waves to San Diego and San Francisco. When the island of Krakatoa in Southeast Asia blew up in 1883, a tsunami with fifteen-meter waves destroyed 165 villages along the Java and Sumatra coasts.[16] Thirty-five thousand people perished along the Sunda Strait coastline alone. As the Lisbon disaster, the Indian Ocean tsunami of 2008, and the Japanese earthquake and tsunami of 2011 remind us (see chapter 10), great tsunamis ravage low-lying shores and raze entire communities with devastating force. The hazard is far greater today than it was in 1755, when Lisbon had 200,000 inhabitants, and the largest city in the world, Beijing, about a million. Today, tens of millions of us are crowded in cities and towns a few meters above sea level. Lisbon alone has 547,000 inhabitants—and rising.

EXTREME WEATHER EVENTS come in many forms—blanketing snowstorms, tornadoes, torrential rainfall, and long-enduring droughts, to mention only a few. However, the most dangerous are hurricanes and tropical cyclones, which generate not only powerful winds and sheets of rain, but also violent sea surges. The infamous Hurricane Katrina, which devastated New Orleans in 2005, alerted us forcibly to the dangers of exceptional storms along low coasts besieged by subsidence and rising sea levels. As we describe in chapter 13, much of the damage and loss of life came not from the hurricane-force winds and rain, but from the sea surge and high tides that followed on the storm. Raging waters swept ashore and carried away entire parishes and massive artificial levees that protected low-lying parts of New Orleans.

Hurricanes like Katrina generate sea surges by the wind blowing directly toward shore and pushing water up onto the land. This is what devastated the Mississippi delta in 2005 and Galveston, Texas, in September 1900, when a hurricane-generated surge flooded the city streets to a depth of at least six meters, destroying thirty-five hundred buildings and killing over six thousand people. Since the Galveston disaster, improved early warning systems, seawalls, and stronger buildings have reduced casualties in better-developed parts of the world, but rising urban populations and the complex and expensive logistics of warning, evacuation, and recovery make it increasingly difficult to avoid truly catastrophic human and material destruction.

Tropical cyclones are a major hazard in many parts of the world, notably in the western Pacific and the Bay of Bengal. Low-lying Bangladesh is basically a huge river delta at the head of the bay, where tropical cyclones breed, cover large areas, and move northward into the funnel created by the coasts on either side of the ocean.

We are already reaping a whirlwind of vicious assaults by an ocean that once lay 122 meters below today's threatened shorelines. Billions of people are at risk from an attacking sea. Our future will be challenging, even before one factors in the ever-present threat of earthquakes and tsunamis. As history shows us, our vulnerability to an encroaching and often aggressive ocean has increased exponentially, especially since the rapid population growth of the Industrial Revolution. While as recently as eight thousand years ago, only a few tens of thousands of people lived at risk from rising waters—and they could adapt readily by upping stakes and moving—today millions of us live in imminent danger from the attacking ocean and from the savage weather events that await in a warmer future.

Millennia of Dramatic Change

There is one knows not what sweet mystery about this
sea, whose gently awful stirrings seem to speak of
some hidden soul beneath.

—Herman Melville, 1851

The nine millennia between fifteen thousand and six thousand years
ago saw complex human adjustments to rising seas, but also witnessed
major shifts in people's day-to-day lives. Ancient human societies living
by seacoasts and lakes focused heavily on fishing, sea mammal hunting,
and fowling. Many groups lived in such food-rich environments like
northern Europe, Southeast Asia, and parts of the South African coastline, dwelling at the same locations for generations. In many parts of the
world, large Ice Age animals became extinct. People turned to smaller
game and to plant foods for much of their diet as the climate changed
the world around them.

Nowhere were the environmental changes more profound than in
northern Europe. For tens of thousands of years, gigantic ice sheets
had covered much of the north. Fifteen thousand years ago, the English
Channel was little more than a large estuary. During the rapid warming
that occurred after the glacial maximum, the sparse hunter-gatherer
societies of northern Europe had to contend with staggering environmental changes. High-tech science has revealed a sunken Ice Age landscape

under the waters of the southern North Sea. Here, low-lying coastlines changed significantly from one generation to the next as the sea attacked the land.

No one knows how many people lived in northern Europe, so intelligent guesstimates are in order. Eight thousand years ago, after the thousand-year Younger Dryas cold snap, perhaps as few as two thousand to three thousand people called northern Europe their home, most of them in the Low Countries and in the now-submerged lowlands of the southern North Sea.[1] Population densities increased reasonably rapidly as warming resumed, to conceivably as many as twenty thousand, if one adds central and northern Scandinavia to the equation. Even the largest communities, situated near bountiful fisheries and lush wetlands, would not have supported more than a few dozen people, probably significantly fewer. People were so thin on the ground at first that moving in the face of encroaching seawater or shifting camp to higher ground was a routine practiced with effortless familiarity in a little-known sunken North Sea world known to scientists as Doggerland.

From Doggerland, we travel to the Euxine Lake, now the Black Sea, to witness the dramatic ravages of an encroaching ocean, which may have had life-changing effects on European societies of the day. Here, farming villages and fertile agricultural land vanished under rapidly rising seawater, perhaps within a few weeks or months. From the threshold of the now-flooded Euxine, our journey takes us to the Nile delta, then to Mesopotamia, the Land Between the Rivers. Both areas witnessed major environmental changes that affected hunters and subsistence farmers, and then growing cities and nascent civilizations. As we will see in later chapters, the same low-lying environments are still a magnet of human settlement in the twenty-first century.

Doggerland

THE NORTH SEA IS SHALLOW, vicious, and unrelenting in its sudden weather shifts. Steep waves assault you on every side even in moderate winds. The fogs are dense and notorious; thick haze is a way of life. I once sailed from Den Helder in the northern Netherlands bound for the Dover Strait. We had sat in port for four days waiting out persistent southwesterly gales. At last they moderated and the wind shifted to the northwest. Full sail and a departure on top of the tide: We had smooth seas and a pleasant, favorable breeze. The idyll lasted for fifty kilometers. Our barometer tumbled four tenths in as many hours, but we were clear of off-lying dangers and kept going. By midafternoon, we were well reefed down and running at full speed before a forty-knot gale. Even with the storm blowing from astern, the seas were violent and cooking hot soup in the galley was an acrobatic exercise. Six hours later, having dodged a supertanker anchored off Rotterdam, we were becalmed once more. Yet the weather forecast spoke of southerly winds of thirty-five knots or so, moderate to rough seas, and squally showers. Fortunately the southerly winds veered to the northwest and we sailed on before long.

"REMEMBER THAT THIS was once dry land," one of the crew remarked as we once slatted back and forth on the Dogger Bank and he gulped coffee laced with spray. "Hard to believe, isn't it?" I was the only archaeologist aboard and I must confess that I'd forgotten that we were sailing over a seabed that had been dry, albeit marshy, land only eight thousand years ago.

In 1931, a British trawler, the *Colinda,* was working the Leman and Ower Banks in the southern North Sea. The trawler men cursed when their net brought up lumps of peat, known as "moorlog," from a depth of eighteen meters. Their nets routinely tore on waterlogged wood and mud lumps as they trawled for bottom fish on the once-marshy seabed. But this time a peat block emitted an unfamiliar sound when hit with a shovel. The skipper broke it open. Out fell a beautifully preserved antler harpoon. Intrigued, the skipper brought the find back to port. The unexpected artifact found its way to the British Museum, which identified it as a hunting weapon used commonly some eight thousand years ago. But the staff didn't want it, because they already possessed three found on dry land. Eventually, the *Colinda*'s find ended up in the Norwich Castle Museum in East Anglia. On February 29, 1932, the members of the Prehistoric Society of East Anglia admired the harpoon, which was identical to such weapons found along the shores of the Baltic in Denmark on the other side of the North Sea. But how had it traveled so far from land? Had some hunter in a canoe dropped it while on a deep-sea fishing expedition or while crossing to the Continent? Or had people crossed from the continent to Britain at a time when a low-lying plain joined northwestern Europe to the higher ground of southern England?[1]

The *Colinda*'s harpoon was by no means the first discovery to be dredged from the North Sea bed. Throughout the nineteenth century, oyster dredgers working the shallow waters off eastern England brought up the bones of extinct animals in their nets. As fishing technology improved and the trawlers moved into deeper water, finds from the Dogger Bank proliferated.[2] This famous shoal lies about a hundred kilometers off the English coast, rising about forty-five meters above the seabed and forming a 17,600-square-kilometer subsurface plateau about 260 kilometers from north to south. Early trawling scoured the surface of the plateau of the bones of such animals as bear, bison, horse, mammoth, and deer. Some of the animals like the long-extinct mammoth and woolly rhinoceros obviously came from the Ice Age, while many were from more recent, undisturbed levels uprooted by the trawls. But

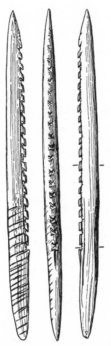

Figure 2.1 *The Leman and Ower harpoon, dredged from the floor of the North Sea. Two thirds full size. Courtesy: The Prehistoric Society.*

the Leman and Ower harpoon added a human dimension to a geological and paleontological mystery that had been around for decades. Here was a human artifact in the middle of the sea.

The mystery began to unravel when Victorian geologist Clement Reid, who worked for the Geological Survey, became fascinated by the stunted tree stumps from long-vanished forests uncovered at low tide along the eastern English coast.[3] Blackened oak stumps, hazel, and alder trees, and the bones of extinct animals came from the once-flourishing forests. None other than the diarist Samuel Pepys had noted ancient hazel fragments preserved in the filthy mud of Thames dockyards. How had such dense forests vanished under the waves? Inevitably, in a devout age, even expert scientific observers assumed the trees had succumbed to the biblical flood. And equally inevitably, they became known as Noah's Woods. Everyone except Clement Reid shunned the half-submerged forests, mantled as they were by glutinous mud and filth, and awash at high tide. As one expert wrote, they "belong to the province of geology, and the geologist remarks that they are too modern to be worth his attention."[4]

Reid was the first scientist to gaze under the heaving surface of the North Sea. He wrote of an extensive alluvial plain that had once covered the entire southern North Sea, about thirty-six meters below modern sea level. Judging from the plant remains, much of this submerged land was marsh and fen, protected by extensive sand dunes and crisscrossed by major rivers linked to the modern-day Rhine and Thames. Clement Reid never named his submerged land. Nevertheless, he published his findings in a short essay, *Submerged Forests*, which appeared in 1913, described in his obituary as "a delightful little book." "Nothing," he said, "but a change of sea level could account for the coastal forests."[5] He noticed that the Humber and Thames valleys cut about eighteen meters below modern ground level and found a similar horizon over wider areas. Perhaps a plain that lay far below today's land surface and sea level flanked these large rivers. Boldly, Reid extrapolated his sunken ground surface to a submerged plain that covered the entire southern North Sea.[6]

Clement Reid died in 1916, leaving a drowned, anonymous land behind him and a prevailing scientific opinion that the southern North Sea had been a vast fen, not a useful landscape, but one that people crossed on their way to higher ground. There matters remained for years, despite the development of a new scientific method for studying marsh deposits, developed in Scandinavia during Reid's closing years. Palynology, otherwise known as pollen analysis, studied minute fossil pollens preserved in marshes and peats. Two Cambridge botanists, Harry and Margaret Godwin, persuaded the skipper of the *Colinda* to return to the Leman and Ower, where they obtained samples of moorlog for pollen analysis. (The original block that contained the harpoon had been thrown overboard.) The Godwins soon showed that the moorlog came from a deposit that had formed in freshwater. So the *Colinda*'s harpoon had not been lost at sea but dropped on land. Cambridge archaeologist Grahame Clark, a friend of the Godwins, wrote a seminal book entitled

The Mesolithic Settlement of Northern Europe in 1936, in which he not only described rich post–Ice Age hunter-gatherer societies in Scandinavia but also waxed lyrical about the favorable prehistoric environments that now lay under the North Sea—a land bridge to higher ground.[7]

By the 1990s, geologists and geomorphologists had collected vast reservoirs of information about the North Sea bed as part of oil exploration. They established that sea levels had risen no less than 120 meters since the last Ice Age maximum, inundating an area larger than the entire United Kingdom. However, their research was short on topographic detail, the kind of information that archaeologists need to survey an area for potential sites.

At this point, the archaeologists sat up and took notice. In 1998, one of them, Bryony Coles of the University of Exeter, published an important paper in which she summarized all the evidence that bore on the ancient North Sea landscape, from about twenty thousand years ago up to the final disappearance of the landscape in about 5500 B.C.E.[8] She called the sunken landscape Doggerland and produced a series of maps based on bathymetric contours that chronicled the slowly vanishing land surface. At the height of the last glaciation, the North Sea did not exist. Land extended from the Shetland Islands to Norway. As sea levels rose, the coastline slowly receded, forming a large inlet, with the Dogger Hills on the eastern side of the narrowing sea. Eventually, as the sea rose ever higher, the Dogger became an island before itself vanishing beneath the waves. Coles's hypothetical maps sparked a new generation of research, in the belief that what had happened on higher ground, especially in Britain, might well have been peripheral to what had happened below modern sea level. However, the murky and turbulent waters of the North Sea were seemingly as inaccessible to archaeologists in 2000 as they had been in Clement Reid's day.

FIFTEEN THOUSAND YEARS ago, the ice sheets that mantled parts of Britain and much of Scandinavia were in full retreat. The permafrosted arctic desert with its cold-tolerant herbs and shrubs that covered much of central and western Europe gave way to more wooded landscapes.

The constant winds carried seeds northward, as did birds and other migrating animals. Fifteen hundred years later, extensive birch, pine, and poplar forests covered Britain, northern Germany, and much of southern Scandinavia. By this time, rising sea levels had reduced Doggerland to a land of extensive coastlines and estuaries.

The rhythms of people's lives changed to accommodate new realities.[9] Some of the greatest environmental changes occurred along northern Europe's coastlines. Rising sea levels inundated continental shelves, filled estuaries, and turned many once rapidly flowing rivers into sluggish streams. The encroaching ocean created rich shoreline environments and wetlands, where fish and waterfowl abounded and shellfish could be gathered by the thousand. Plant foods like edible seaweed were often plentiful. Some of the richest of these low-lying coastal and estuarine environments lay in Doggerland and along the coasts of a large glacial lake that was to become the Baltic Sea.

As the Scandinavian ice sheet retreated, an ice lake had formed along its southern rim, dammed by low hills along what are now the north German and Polish coasts. When the weight of the ice lightened, the land rose faster than the sea. Cycles of cooling and warming changed the configuration of what was sometimes part of the ocean, at others a lake. After the Younger Dryas cold event ended in about 9500 B.C.E., renewed warming opened up an outlet through central Sweden, while a land bridge formed over what is now the Oresund between Denmark and Sweden. A brackish sea survived for about twenty-five hundred years before the rising land again turned it into another lake. Both were ancestors of the Baltic Sea. Around 5500 B.C.E., just as the North Sea flooded the last of Doggerland, the sea once again broke through and formed the direct ancestor of the modern brackish Baltic Sea.[10] The Baltic's changing, dynamic landscapes proved to be a magnet for human settlement as early as 7000 B.C.E., perhaps earlier.

The fishermen stand motionless in the cold, shallow water, barbed spears poised at the ready. A man's toes feel the mud, sense the imperceptible movements of a flounder on the bottom. A quick stab with the spear and a wriggling fish comes to the surface, firmly impaled on the spear barbs. The fisher quickly wades ashore and kills his catch with a wooden club.

Doggerland's low-lying shorelines with their creeks, sandbanks, and wetlands, and those of the Baltic, were a veritable garden of eden for the hunting bands that camped along northern coasts. They were fishers and fowlers, adept at collecting plant foods, falling back on shell-fish when other foods were in short supply. Huge piles of discarded mollusks—shell middens—lie at strategic locations on Danish and Swedish coasts and must also lurk underwater in the North Sea. Food supplies were so abundant that populations rose steadily over many generations. Everyone dwelled at first in small encampments, shifting locations within hunting territories where there was plenty of room for everyone. Their possessions were the simplest, most of them fashioned from antler, bone, and wood, dugout canoes hollowed from logs with flint axes, wooden fishing spears, nets, and fish traps. Two weapons were all-important: the barbed fish spear, often with two or three antler or bone heads (like the Leman and Ower point) that impaled fish like pike or salmon with its prongs, and the bow and arrow. The hunters shot birds in flight with arrows tipped with tiny, lethally sharp stone tips. They also hunted deer and a formidable prey, the wild ox.

So plentiful were food supplies of all kinds that many bands lived within very circumscribed territories. But the richness of the Baltic environments must have paled beside those of Doggerland's low-relief world of estuaries and marshes, low ridges, shallow valleys, and extensive wetlands. It was here, in a now-submerged world, that the densest hunting populations must have thrived, adjusting gradually to a radically changing world.

WHAT WAS DOGGERLAND really like? What varied landscapes once lay under the North Sea? A major breakthrough came when archaeologists Vincent Gaffney and Simon Fitch of Birmingham University dug into seismic data collected by the oil industry.[11] Oil exploration aimed at deep sediments, not at the shallow, much more geologically recent deposits close to the seafloor where human artifacts might be found. Petroleum Geo-Services (PGS), a geophysical prospection company, generously gave the archaeologists access to six thousand square kilometers of seismic data from the Dogger Bank for a pilot study. Less than a month later, the shadowy course of a hitherto unknown river winding over the Dogger Bank for forty kilometers appeared on their computer screens. So successful were the preliminary results that PGS donated no less than twenty-three thousand square kilometers of three-dimensional seismic data from English territorial waters in the southern North Sea, a survey area slightly smaller than Belgium. A team of archaeologists, geologists, and palaeoenvironmentalists spent eighteen months exploring an unknown prehistoric land.

The seismic data provided information on rivers, estuaries, and lakes, as well as salt marshes and coastlines. The research located about sixteen hundred kilometers of river channels and twenty-four lakes or marshes, the largest of which covered more than three hundred square kilometers. This was a very watery landscape. At its heart lay an immense bathymetric depression known today as the Outer Silver Pit, perhaps an estuary or lake off the Dogger Bank, which has been mapped over seventeen hundred square kilometers. Great sandbanks at the eastern end hint that it eventually became an estuary for two major rivers or channels entering from the east and southeast. As sea levels rose, the river mouths may have become marshes, one of which, an extensive salt marsh, must have been a major resource for local hunters, with its abundant fish, bird, and plant life.

At least three rivers flowed from the slightly higher ground that is now the Dogger Bank. They meandered through wide valleys, fed by numerous small streams. Judging from what we know about contemporary valleys on higher ground, watercourses large and small flowed through a network of channels quite unlike the more well-regulated course of the Thames or Rhine Rivers of today. The channels were unstable, changing constantly with sudden floods and course shifts that helped form mazes of back channels, creeks, and swamps.

It was a monotonous, water-filled landscape, one might think, but to the hunting bands who lived there, the environment would offer rich

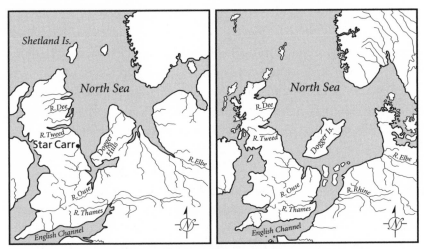

Figure 2.2 *The main features of Doggerland ca. 10,000 years ago
(left) and ca. 6,500 years ago (right).*

potential and all kinds of subtle landmarks for finding one's way around.
An outcrop of white sand, a clump of trees overhanging a small pond,
deep reedbeds where a hunting platform could remain undetected—all
were inconspicuous markers. So were shallow lakes where a hunter could
wear a decoy headdress, swim quietly among sitting ducks, and pluck
them from below the surface. Deer, pigs, and numerous small mammals,
and also myriad waterfowl, congregated along riverbanks, by lakes, and
in marshes. Fish must have dominated many diets. Salmon runs alone
would have sustained some bands for months on end. Plant resources
were plentiful—not only with reeds to make baskets, fish traps, and huts,
but also as food sources with a broad range of seasonal berries and
fruit, hazelnuts, and acorns as forest spread across the landscape. Each
family, each group, would have passed knowledge of the food supplies
that changed from season to season from one generation to the next in
an endless mnemonic that mapped the landscape across hunting territories large and small.

Unfortunately, seismic data doesn't provide information on archaeological sites, which are inaccessible far below sea level. Gaffney and his
colleagues have ranked various features located in Doggerland as to
their potential for archaeological discovery. Highest ranked, as one

would expect from experience with sites on higher ground, are lake and marsh areas, also coasts. One day a detailed map might provide enough information for a more thorough exploration of the landscape by ships using core borers to sample subsurface deposits in the hope they will yield proxy traces of human activity, such as pollen grains documenting deliberate forest clearance or humanly set fires to foster new growth. Meanwhile, we have to extrapolate from known sites on higher ground.

IN ABOUT 8500 B.C.E., a hunting band camped on the shores of a small glacial lake in what is now northeastern England. Dense reedbeds surrounded the water; birch woodland pressed to water's edge. The resulting waterlogged archaeological site, known as Star Carr, preserves an unusually fine-grained chronicle of hunter-gatherer life in a landscape that must also have flourished in Doggerland, close to the east. Grahame Clark of Leman and Ower fame excavated Star Carr between 1949 and 1951.[12] He wrote of a small winter hunting platform deep in lakeside reeds used by red deer hunters. Clark described a remarkable series of bone and wooden tools, including elk antler mattock heads (one with the tip of its wooden handle still in place), a wooden canoe paddle, awls, and even bark rolls and lumps of moss used for fire lighting. Several excavators have returned to Star Carr since Clark's day, using all the resources of today's high-tech archaeology to reinterpret the site.

Clark had focused on the waterlogged portions of the site. A new generation of researchers has traced drier areas of Star Carr and revealed what was actually a far larger settlement that once extended over some 120 meters of the shoreline. By dating minute charcoal particles, researcher Petra Dark was able to show that the inhabitants had burnt off the reeds repeatedly, perhaps to provide a better view of the lake and surrounding terrain, and also a good canoe-landing place. Such repeated burnings occurred for some 130 years. There was then a century with no burning, followed by another eight decades of more firing. Star Carr's hunters took many ten- to eleven-month-old deer, which would have been killed between March and April. Reed samples and aspen buds also date to the same time of year, with Star Carr being occupied between March and June or July each year. Recent excavations have revealed a circular house with at least eighteen wooden posts and a sunken floor that was 3.5 meters wide, and also a wooden platform made of split and hewn timbers. This substantial dwelling may reflect more permanent settlement than is often assumed for the hunters of 10,500 years ago. Whether there were similar important, and repeatedly visited, locations like Star Carr in Doggerland is of course unknown. But there is no reason why there should not have been, for environmental conditions were much the same over a vast area of what is now the southern North Sea.

THE SOFT *SCRAPE, SCRAPE* sound of flint scrapers cleaning a pegged-out deerskin would have greeted a visitor to a Doggerland hunting camp in 7000 B.C.E. In this cold, often rainy world, skin cloaks and fur garments were essential for young and old. Like all hunters, the women would have made use of a variety of skins for different garments, perhaps tough seal hide for boots, soft rabbit fur for decoration and baby clothes. A small cluster of grass huts, perhaps a central hearth, dogs, children, and a smell of decaying fish—the scene would have been the same at dozens of such camps. Weeks, even months, could go by without the band seeing anyone else. If they did, it was probably an event, a matter of a cautious approach and careful exploration of potential kin relationships. The population of Doggerland would never have been large, dwarfed by a seemingly endless world of dunes, lakes, wetlands, and woodland. You would have glimpsed humans but rarely, usually on slightly higher, better-drained ground, or camped among reeds or by lakes or rivers. A wisp of woodsmoke from a hearth, the soft chipping sound of a stone adze carving a dugout canoe, women scraping a deerskin pegged out on damp ground: Your impressions would have been fleeting, dwarfed by the marshy landscape. In a few locations, dark smoke would rise from dry reedbeds in spring, as the hunters set fire to undergrowth to gain better access to lakes and streams, and to foster new plant growth for game.

Figure 2.3 *Artist's reconstruction of life in a Doggerland hunting camp, ca. 7000 B.C.E.* © *Bob Brobbel.*

Everywhere in this sunken land, people lived at the mercy of the rising ocean, for, even within generational memory, seawater would inundate mussel beds and beaches where their ancestors had once gathered food. Rivers would alter course without warning; severe winter storms would bring storm surges and exceptionally high tides; sand dunes would encroach on favored wetlands. The water was inexorable, ever intrusive, never retreating, to the point that high tide marks became low tide levels within a few years. Once-conspicuous ridges became islands surrounded by shallow water. As the waters rose, so each band adjusted easily to more restricted territories or moved to higher ground.

Doggerland became more and more of a watery world, dotted with ever-larger lakes and increasingly sluggish rivers that overflowed their banks, their gradients reduced by higher sea levels. Century by century, the wetlands vanished as the North Sea gradually became a complex archipelago of islands. The largest was the Dogger Hills, which remained above water long after the rest of Doggerland had vanished. By that time, most people had moved to higher ground in what is now the Low Countries or Britain. Increasingly, Doggerland must have remained only as a folk memory, as a realm of vanished lands once inhabited by revered ancestors. The gray, muddy waters with their violent storms and short, steep waves became the home of the dead, of those who came before. As for the dead themselves, at least some of them were buried on islands amid the waters, if analogies from higher ground are to be trusted. As the waters rose and the islands vanished, so the dead entered the place of those who went before, remembered only in oral tradition and folk memory. We know of such burial places from Somerset in southwestern Britain, where local hunters buried their dead on a low island amid the wetlands of what are called the Somerset Levels.

By about 5500 B.C.E., after ten thousand years, Doggerland had become the North Sea. The last of the Dogger Hills had vanished. The former inhabitants of what was now ocean had long before moved to higher ground, but many of them remained along still gradually flooding coastlines and estuaries nearby. The densest populations were along Baltic shores and especially in Denmark, where hundreds of abandoned campsites lie both at higher elevations and in shallow water. Such waterlogged locations have proved a treasure trove of rarely preserved archaeological finds such as fishing nets and traps, antler, bone, and wood spears, even paddles and rare dugout canoes. At the Ulkestrup site in Denmark, people lived in large huts with bark and wood floors on a peat island nestled in a swamp by a lake. One hut lay close to where canoes were once moored. These were people who thrived, as Doggerland refugees did, near shallow water and wetlands, where they preyed on fish, mollusks, and waterfowl.

As warming continued, populations rose in the most favored coastal areas. Hunting territories became progressively smaller. For example, the Segebro settlement near the southwestern Swedish coast now lies in brackish water. The small encampment close to the water once covered fifty by twenty-five meters. The inhabitants lived there year-round, but mainly in spring and summer. Carbon isotope analysis of their bones shows that fish and sea mammals made up 60 percent of the diet. Nevertheless, they enjoyed an omnivorous diet. The remains of no less than sixty-six animal species lay in the site.

By 4600 B.C.E., about a thousand years after the final inundation of Doggerland, many hunting bands lived in more or less sedentary settlements, often occupied for months at a time. Territories were smaller, the boundaries more closely defined, for prime hunting grounds had shrunk in the face of rising seawater. With much greater population densities in the richest and most diverse environments, local societies became more complex, their artifacts and art styles more diverse, as if cultural differences were assuming greater importance. Differences there may have been, but there are signs that neighboring groups remained in contact, trading such commodities as tool-making stone over considerable distances. At the same time, the food quest intensified in the face of greater crowding.

Such an intensification of the food quest was inevitable in a changing world of increasing competition for valuable food resources. Neighbors eyed nearby oyster beds or fishing grounds; hunters quarreled over who had the right to take migrating waterfowl resting on small lakes in spring or fall. The ownership of salmon runs and eel-rich bays must have led to feuds and occasional violence. No longer could people move away and explore vacant land. Now their neighbors were close by and territorial boundaries became increasingly sacrosanct as people stayed in more or less permanent settlements anchored closely to their ancestral lands. Sometimes there may not have been enough food to go around.

With sedentary settlements, many communities now buried their dead in cemeteries. In the cemetery at Vedbaek Bogebakken in Denmark, used around 5000 B.C.E., twenty-two people of different ages lie in shallow graves.[13] Some perished after violent deaths, one lying with a bone point embedded in his throat. A few burials will never provide a full or accurate picture of violence along these changing shores, but the signs of stress—of increasing crowding—are there in a world where the sea pressed aggressively on the land. There was far less hunting territory to go around. For thousands of years, the hunters of Doggerland had moved constantly in the face of rising seas. Now their primordial homeland had vanished to the distant recesses of human memory, but the still-attacking sea compelled their remote descendants to settle in one place, to exploit a far wider range of foods, and moved some of them to competition and violence.

For thousands of years, a sparse population of hunters and foragers had engaged in an intricate minuet with the ocean that lapped at their proverbial doorsteps. Even a large band would stay for only a few years at a favored water's-edge location before moving elsewhere as dry land and marsh vanished underwater. Generational memory and cherished oral traditions would have underlain a way of life defined by movement, not environmental change. Now the equations of daily life were changing. People lived in larger settlements occupied for decades, even centuries. They could still respond to an attacking ocean by moving, but the process was more intricate and thwart with complex social realities, many of them defined by ancestral ties to hunting territories and fishing grounds. In these changes, we can detect the first signs of a pervasive human vulnerability to sea level change in a slightly more crowded world.

Euxine and Ta-Mehu

THE SEA OF MARMARA, Turkey, 5500 B.C.E. A powerful winter storm sweeps up the constricted strait. Great breakers thunder on a low-lying sandy beach at the upper end. Spray flies. The water level in the sound rises with the mounting storm. High surf flows up small gullies farther ashore, then recedes, eroding slender defiles and forming inlets. When the wind drops and the water recedes, the ridge, an ancient pathway that separates the Mediterranean from the great lake far below, is narrower than ever before. Each spring local farmers cross the shrinking berm with their flocks toward the higher ground to the west. Each fall, they return, following paths used since time immemorial. As their beasts crowd the track, the shepherds glance down at the ruffled blue waters of the lake far below, where lush water meadows nourish winter pastures.

Year after year, the defile becomes narrower. Storm after storm, the erosion persists as sea levels rise almost imperceptibly. Steep waves eat away at the beach and sandy cliffs. Then, inevitably, at the height of a major storm, surging water from a set of steep breakers bursts through to the slope on the other side. The damage is done, a channel formed. Wave after wave deepens the furrow, until seawater courses through even in calm weather. Now the inexorable force of gravity takes over. The inlet becomes a saltwater outlet that flows deeper and wider. Within days, the gully becomes a torrent, then a rushing waterfall that carries water flowing at over ninety kilometers an hour. The deeper the water cuts into the berm, the faster it flows. Modern-day estimates tell us that enough water passed through each day to flood Manhattan to a depth of almost a kilometer. Within weeks, the Euxine Lake rises rapidly, perhaps as much as fifteen centimeters a day. After two years of inflow, the Euxine Lake filled to the same level as the Mediterranean and became a brackish ocean, now known as the Black Sea.

DURING THE LATE Ice Age, the Mediterranean (sometimes called the Middle Sea) had been a chilly, much smaller ocean, often swept by vicious northerly winds that brought subzero temperatures and bitter cold. For the most part, the sea level rise had been steady and unspectacular, bringing changes that impacted coastal peoples on, at most, a generational timescale. The flooding of the Euxine Lake was a geological event on a different, potentially catastrophic scale, far more life threatening than the inundation of Doggerland. We know of the flooding from deep-sea cores that chronicle the changing salinity of the water from tiny diatoms and other clues. We also know that the inundation originated in happenings on the far side of the Atlantic, when the great Laurentide ice sheet covering northern Canada finally collapsed. The ice had retreated rapidly after fifteen thousand years ago, as global sea levels began rising in the face of rising temperatures. By about 6200 B.C.E., huge meltwater accumulations finally undermined the shrinking ice sheet. The Laurentide imploded. A massive outflow of meltwater from a large glacial lake, Lake Agassiz, cascaded into the North Atlantic and the Gulf of Mexico. Sea temperatures cooled rapidly. The great ocean conveyor belt that drives the Gulf Stream in the North Atlantic slowed.[1] Like a light switch, the slowing circulation tamed the moist westerlies that brought rainfall to Europe and the eastern Mediterranean. The Balkans and eastern Mediterranean suffered under severe droughts, so much so that many farming societies abandoned their villages as lakes dried up and rivers and streams withered in the cool, arid conditions that lasted for some four centuries.

This was not the first cold snap to affect Europe and the Middle East after the Ice Age. A similar episode, the Younger Dryas, also caused by melting ice, had at least partially shut down the Gulf Stream six thousand years earlier and brought cold and drought that endured far longer, for at least ten centuries. That long drought cycle had seen hunting societies in the Middle East switch from a close dependence on nuts and wild grasses to cultivating cereals and domesticating animals. The switchover to farming, an entirely logical response to shrinking wild grass stands, may not have been easy, but it was to revolutionize human history. As the cold snap and accompanying droughts eased, the new farming economies had spread rapidly throughout the Middle East and along the eastern Mediterranean coast.[2] The farmers sought out easily cultivable fertile soils, many of them on river floodplains and near lakes. There they thrived until dry, cold conditions returned, when persistent droughts anchored many villages to rare, permanent water supplies. Once such refuge would have been the shores of the Euxine Lake, formed in a huge basin by meltwater from northern ice sheets.

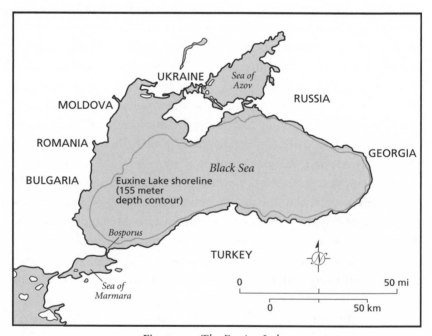

Figure 3.1 *The Euxine Lake.*

The Euxine Lake lay some nine hundred meters below the Anatolian Plateau in what is now Turkey to the south, where farmers had settled many centuries earlier. Perhaps the lake basin formed a large oasis. Temperatures around its shores would have been considerably warmer during the cold centuries, water plentiful, and the soils fertile. Pollen grains from deep-sea cores show that the coastal plains around the lake supported grassland and steppe. Farming here did not require large-scale forest clearance, a challenge for people with stone axes. Most likely a denser concentration of people lived around the Euxine than elsewhere. They would have dwelled in small villages, clusters of mud-brick houses linked by small alleyways, surrounded by patchworks of fields and pastures. There may have been occasional larger communities, especially in locations close to sources of such valuable commodities as tool-making stone. Like their distant contemporaries throughout the Middle East, they would have had close ties to land that had been farmed for generations, since the time of their ancestors.

Sedentary farmers these people might have been, but they also relied on hunting, fishing, and foraging during lean months, living as they did in less crowded landscapes than those of later times. Their agriculture depended on cultivating carefully selected patches of lighter, easily turned soils, which meant that extensive tracts of uncleared landscape surrounded their settlements, an edible landscape rich in wild plants where game abounded. This gave them flexibility that enabled them to subsist off foods other than their crops and stocks of stored cereal grain for considerable periods of time. Then, abruptly, climate change unsettled everything.

Around 5800 B.C.E., the effects of the Laurentide meltdown receded as the Gulf Stream resumed its normal circulation. Warmer, westerly airflows resumed over the Mediterranean; high pressure settled over the Azores. Persistent westerly winds caused temperatures to rise over Europe, leading to a "climatic optimum" that lasted for two thousand years. Before 6000 B.C.E., farming societies had already moved from the Aegean region onto the Great Hungarian Plain and into the Danube River Basin. They prospered in the milder conditions, so much so that people in northern Greece and what is now Bulgaria dwelled at the same locations for many centuries.[3]

The same warm conditions caused sea levels throughout the Mediterranean to rise once more in response to the mass of water added to the ocean by the collapse of the Laurentide ice sheet. During the cold centuries, the Mediterranean was about fifteen meters below modern levels. By about 5600 B.C.E., rising seawater was lapping at the natural berm that separated the Sea of Marmara from the glacial Euxine Lake, which lay about 150 meters lower on the other side. The inevitable then transpired: The berm was breached, and the Euxine Lake became a brackish sea.[4]

The flooding of the Euxine was an environmental change of truly epochal dimensions. But what were the consequences for the farmers living around the lake? We can imagine the sudden confusion as the lake began rising and the water became saltier by the day. The encroaching inundation would have drowned lakeside marshes and swamped growing crops. Canoe landings would have vanished within days; river deltas would have been flooded by muddy water. Thousands of dead fish would have floated in the newly brackish water. Soon afterward, long-established villages with their houses and storage bins would have vanished under the flood. The helpless villagers would have had some time to recover their possessions and empty their grain bins, and to move their herds to higher ground. However much time they had, many communities must have suffered badly from hunger until they were able to establish new villages and clear land away from the rising lake. Psychologically, the moves would have been traumatic, for the lands protected by their revered ancestors would have vanished forever.

The dynamics of life by the lake had changed completely. Villages once well back from the lake now lay at the heads of bays or on exposed shorelines. Many communities must have settled earlier along the innumerable small rivers and streams that led inland to an unknown world of endless forests. They would have moved inland with their herds, dispersing in many directions, following patches of lighter soil and more easily cleared land. Within a few generations, some of these farmers had emerged on the Bulgarian plains and made their way up the Danube River basin into the heart of central Europe, where farmers had never ventured before. In one of the most significant population movements of human history, the descendants of farmers displaced by the rising Black Sea settled a band of easily cultivated glacial soils and river valleys from western Hungary to the North Sea.

IT TOOK ABOUT ten thousand years of warming and rising sea levels to bring the Ice Age Middle Sea close to its current levels. By about 5000 B.C.E., coastlines had basically stabilized near modern heights, although adjustments to higher sea levels continued to persist, especially as rivers' flows became more sluggish and silt brought down by spring floods accumulated on coastal floodplains rather than being carried offshore into deep water. Nowhere in the Mediterranean world were these effects more marked than along the Nile, but, unlike the Euxine flood, the changes in what is now Egypt unfolded over thousands of years.

A century ago, long before the building of the Aswan Dam during the 1960s reduced silt levels, one sensed the presence of the Nile far offshore from the Egyptian coast. Deep layers of silt and river mud colored inshore waters and extended far offshore. That peripatetic Greek traveler Herodotus experienced the approach to the Nile firsthand twenty-five centuries ago: "The physical geography of Egypt is such that, as you approach the country by sea, if you let down a sounding line when you are still a day's journey away from land, you will bring up mud in eleven fathoms [twenty meters] of water. This shows that there is silt this far out."[5] The landfall on a low coastline in front of the prevailing north wind required nice judgment. Herodotus must have watched as his skipper used a lead and line tipped with wax that brought up river silt to calculate his distance offshore.

As they made landfall, Herodotus and his fellow passengers would have gazed on a low-lying, hazy shoreline marked by occasional palm trees and long, sandy beaches. The Nile delta was somewhat of an anticlimax for travelers who had braved the ocean in expectation of ancient wonders. There were no temples or monuments wrought in stone along this featureless coast, apart of course from the marvels of their destination, a growing Alexandria, soon to become one of the great cities of the classical world. Many centuries later, Florence Nightingale of Crimean War nursing fame aptly wrote of the surrounding landscape in 1849: "The dark colour of the waters, the enormous unvarying character of the flat plain, a fringe of date trees here and there, nothing else."[6] One travels by donkey or boat across a level obstacle course of irrigation canals, small fields, and villages. Tales of the Nile with its wondrous summer inundation had spread far and wide by Herodotus's time, but the river lacked a spectacular estuary crowded with oceangoing ships. Instead, it dissipated into an enormous fanlike delta several days' journey upstream from the Mediterranean Coast.

Herodotus thoroughly enjoyed his time in Egypt. His *Histories* masked the Nile valley in a delicious mélange of fact and fantasy. He correctly described ancient Egyptian mummification of the dead, fantasized that the Pyramids of Giza owed their building to pharaoh Khufu's daughter's earnings from prostitution, and recorded all kinds of gossip imparted by priests who lived off gullible tourists. But for all his tall stories, he realized that Egypt and its people depended on the Nile's summer flood and the fertile soils of its great delta. Life in the Nile valley involved a delicate balancing act with an environment marked by unreliable floods and ever-changing sea levels. Herodotus remarked prophetically that should the Nile cease to bring water to the delta, then the Egyptians, with virtually no rainfall, would suffer torments of starvation like the Greeks.

At 6,650 kilometers, the Nile is the longest river in the world, flowing almost directly south to north. In its more northerly reaches it passes through some of the driest landscapes on earth.[7] In Herodotus's day, no one had any idea of the length of the Nile or of the source of its waters. The flood arrived at the First Cataract upstream of Egypt's frontier town Swenet, now Aswan, about 1,127 kilometers south of the Mediterranean, in early summer, seemingly by magic. As the inundation rumbled over the rocky granite outcrops, the ground trembled, so much so that the Egyptians believed that the river originated in a vast underground cavern below the cataract. They made offerings to the rain god Khnum on Abu (Elephantine) Island in the middle of the river, so named because of its associations with the desert ivory trade. Here a Nilometer, a graduated column, allowed priests and officials to estimate the height of the impending flood downstream.

From the cataract, the narrow green Nile floodplain shoots like an arrow across the eastern Sahara Desert. It is as if ancient Egypt were a giant lotus flower. The stalk ran through the narrow valley that is Upper Egypt, then flowered into the blossom, the fan-shaped delta of Lower Egypt, known to the pharaohs as Ta-Mehu, "the flooded land." (The Nile flows from south to north, which means that Upper Egypt is in the south, a confusing label for many people.) From an area near modern-day Cairo (a medieval Islamic city), the delta opens up dramatically into a featureless landscape. It is as if one is entering another country, which is why the ancients spoke of their state as formerly two lands. Close to the Mediterranean shore, the Nile dissolves into once-extensive marshlands and brackish lagoons, what the Egyptologist Toby Wilkinson calls "a shifting landscape, poised between dry land and sea."[8] Here inhabitants once passed from village to village down narrow, reed-swathed defiles where bird life abounded and catfish lingered in the shallows.

The boundary between the Upper and Lower Nile was the area the ancient Egyptians called the Balance of the Two Lands, the place where the Nile splintered from a linear valley into myriad channels. The Balance became the administrative center of Egypt when the first pharaoh, Narmer, unified the Two Lands into a single kingdom around 3000 B.C.E. The pharaohs ruled their domains from nearby Memphis for three millennia. A line of royal pyramids at the edge of the desert extends almost thirty-two kilometers along the edge of the desert cliffs west of the ancient capital. Lest one wonders why the capital lay in the Balance of the Two Lands, consider this: Ta-Mehu has been Egypt's granary for more than five thousand years. Today, the delta's fields, two thirds of the country's habitable lands, provide two thirds of the country's agricultural output. Three thousand years ago, the pharaohs depended on the same acreage to feed their people, so they made sure their

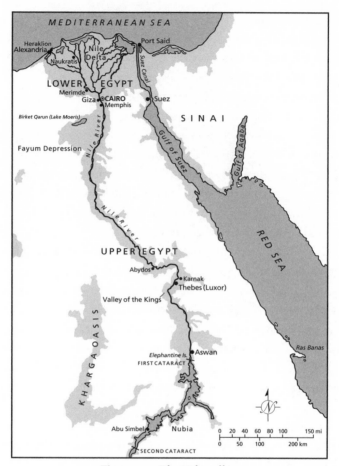

Figure 3.2 *The Nile valley.*

capital was close by. They were also well aware that Ta-Mehu was under siege from the Mediterranean, the northern frontier of their land.

THE NILE DELTA is one of the largest in the world. It covers some 240 kilometers of the Mediterranean shoreline and extends about 160 kilometers upstream to the vicinity of Cairo. This is an arcuate (arc-shaped) delta, like an upended triangle with an arc-like lower edge. The delta has always been vulnerable to rising sea levels, for it is a meeting place between the land and the ocean. Today the outer margins of the delta are eroding in the face of destructive waves driven by the prevailing winds and winter storms of a rising Mediterranean. In places, the coastline is advancing as much as ninety-one horizontal meters a year. By 2025, experts project a sea level rise of thirty centimeters, which will inundate about two hundred square kilometers of agricultural land. The Nile delta is slowly becoming a salty wasteland, yet about half of Egypt's eighty million people live in the general region, with rural populations alone averaging one thousand people per square kilometer.

For all its dense population today, Ta-Mehu is a geologically new land.[9] Eighteen thousand years ago, the late Ice Age Nile flowed into a Mediterranean that was much lower than today. Cores bored into delta deposits and the underlying bedrock tell us the coastline was at least fifty kilometers farther north than today. At the time, the Nile flowed across an alluvial plain dissected by numerous small channels. The faster-flowing river waters of the day deposited little silt on the plain. Had you visited the mouth of the Nile, you would have gazed across a gently undulating near desert, a sharp contrast to the lush floodplain of today. Your feet would have crunched on coarse river gravel brought down by summer floods. The Nile itself would have flowed vigorously down a steeper gradient than today. At flood time, some of the water would spill over into shallow depressions, where stunted grass would grow in the lingering damp. But the landscape would have been bleak and inhospitable, even at the shore, which was devoid of the lagoons and lakes that formed in later times.

After about 7500 B.C.E., sea levels climbed rapidly. An unstable, constantly changing shoreline migrated southward as the Mediterranean flooded broad tracts of coastal lands and deposited marine sand over a wide area. One would still have trodden on gravel rather than fine river silt, despite a shallowing river gradient. The coastline would still have been a desolate waste of sandy beaches and blowing dunes with no standing water except in the river and its side channels. The big change came around 5500 B.C.E., when sea level rise slowed, just as it did in the Persian Gulf. The now much more sluggish Nile ponded in the neighborhood of modern Cairo. Here, at the Balance of the Two Lands, the stalk of the lotus—a relatively narrow valley—now blossomed into a broad delta, the flower of the long plant. As the ponding, and now much slower flowing, Nile deposited enormous quantities of silt along its course, the delta assumed its basically modern configuration. Now dozens of channels large and small carried the river to the sea. At flood time, the inundation spread over the delta, effectively waterlogging much of the land. Meanwhile, the Sebennetic channel of the Nile, in the western part of the delta, transported so much coarse sand to the coast that it formed an extensive sand barrier, a natural fortification against the rising sea. By now, tiny numbers of hunters and foragers must have settled the flat landscape, not so much the flatlands, but the extensive lagoons, lakes, and marshes that formed behind the shoreline. Here, as in southern Mesopotamia, extensive wetlands served as a magnet not only for small numbers of humans, but also for migrating waterfowl on the great Nile valley flyway. Rich in fish and plant foods, and also reeds for housing and canoes, the coastal waterways soon attracted hunters, and, shortly afterward, farmers and herders.

As humans settled on the delta, changes wrought by the river and ocean became more muted. By 4500 B.C.E., the Mediterranean was still about ten meters below modern levels, the river gradient was somewhat steeper than today, and the climate was somewhat wetter. However, sea level rise continued more slowly and the delta's gradient shallowed around 2000 B.C.E. as the climate became significantly drier. No less than five river channels still deposited large quantities of silt along the coast. Two thousand years later, at about the time of Christ, the sea level was only about two meters below modern levels. Extensive wetlands continued to flourish along a coast increasingly shaped by waves and sea level fluctuations. As the climate became drier, so the lagoons and lakes by the coast continued to be a major source of fish, waterfowl, and other foods.

RISING SEA LEVELS and river silt shaped Ta-Mehu and its wetlands, but ultimately it was humans who turned the delta, and all of Egypt, into what one can accurately call an organized oasis. The process took many centuries and continues to this day.

Fifteen thousand years ago, when the Ice Age ended, the entire human population of the Nile valley from the Sudan to the Mediterranean was probably little more than a thousand souls. They lived in small bands, scattered in large hunting territories along the Nile banks. As sea levels rose and the river filled the valley with sediment, hunting populations increased, especially in favored areas, one of which must have been the coastal wetlands of the delta. But the human imprint would have been faint. Smoke from a fire at the edge of the reeds, dogs barking, the occasional fisher spearing catfish in shallow water: the people of the delta relied on a highly portable tool kit and would have moved from place to place within large territories depending on the season of the year.

The blossom of the lotus, the delta, was open to a much wider eastern Mediterranean world, where farming began before 9000 B.C.E. At the time when agriculture began to spread rapidly throughout the Middle East after 8000 B.C.E., the delta was still an arid floodplain with a rapidly changing coastline, which may have made farming a difficult proposition until sea level rise slowed around 5500 B.C.E. and floodwaters and silt nourished the flat and hitherto infertile terrain. Most likely, most hunting bands living near the mouth of the Nile settled close to the wetlands behind the coast, where fish and waterfowl, as well as reeds and plant foods, abounded. The situation was like that in Doggerland and southern Mesopotamia, where the rich food resources of marshes formed by interaction of rivers and a rising ocean created a swath of wetland environments. In such landscapes, whether in Europe or in Mediterranean lands, farming may have begun as a side activity to supplement a diet of game, fish, and plants, with the foods close at hand in the wetlands serving as backup against crop failure.

Farming soon began to play a more prominent role in human existence. The Nile world changed completely within a thousand years. No one knows when farming first came to the isolated world of the Nile valley. We can never prove it, thanks to deep silt accumulations, but most likely Egypt's first farming communities developed close to Ta-Mehu's marshlands in the shifting landscape of the northern delta. At first, there would have been few changes. Every day, fishers standing in canoes would spear fish in quiet lagoons. In winter, the villagers would trap waterfowl or hunt them with bows and arrows as migrants arrived by the thousand. Thus it had been for centuries, and so it continued, except for patches of cleared ground close to small, huddled villages of reed huts, where wheat would ripen in winter. Like the ancestors of the Marsh Arabs in southern Mesopotamia (see chapter 4), such communities must have continued to flourish for many centuries, deep into the time of the pharaohs, people on the margins, protected from the outside world by narrow channels and thick reeds.[10]

Elsewhere farming soon transformed daily life. Fish and game were still important. The farmers went out at dawn after waterfowl with stone-tipped arrows or after fish with spears. Everyone continued to dwell in low reed-and-matting houses partially sunk into the ground. The small fields lay close to the water, the crops harvested with flint-bladed sickles in spring, the grain stored in subterranean pits, where archaeologists found seeds thousands of years later. These humble farming settlements, whether along the great river or around the wetlands and lakes of the deltas, were the forebears of the thousands of agricultural villages that supported the elaborate panoply of ancient Egyptian civilization in later times.

Some of the delta's shadowy wetland communities thrived for many centuries, but other farmers flourished in the delta away from the marshes, growing crops in the months after the inundation. We have a fleeting portrait of one such village named Merimde Beni-salame, northwest of Cairo, occupied as early as 4800 B.C.E., where farmers dwelled for as long as a thousand years.[11] Like the earliest farmers in Mesopotamia (see chapter 4), the Merimde people relied heavily on local raw materials, especially reeds. They lived in small wattle-and-reed huts with oval floor plans. We can imagine a huddled settlement of weathered, circular dwellings set near marshy terrain, surrounded by small fields and tucked among tall reeds, standing on slightly higher ground that remained dry during the inundation. In the hot and humid late summer, the waters would recede, the stubble from the previous crop would appear in the fields. Standing ankle-deep in the floodwaters, men and women would turn over the rich black soil and the freshly deposited silt with digging sticks and stone-bladed hoes. If the inundation was bountiful, green shoots of the new wheat crop would appear in the damp earth a few weeks after planting.

There were dozens of small farming villages like Merimde across the delta by 4000 B.C.E., places where life remained unchanged for many generations, governed by the waters of the Nile. Come late summer, floodwaters from far upstream would reach the delta, overflow from braided river channels, and settle over harvested fields. As the inundation receded, fine silt suspended in the water would descend on the cleared land. Most years, the villagers would divert the water from plot to plot with small dikes, turning the wet ground with digging sticks and stone-bladed houses for the new crop. Then came planting, growth, and, months later, harvest, an endless cycle of backbreaking labor even in good flood years. The annual waters flushed out salt from the ground; the river silt fertilized the soil and helped keep the ocean at bay.

Each village was a seemingly isolated community, but ancient kin ties and complex marriage and social relationships linked settlements near and far. Everyone living in the delta must have been aware of a wider world far beyond the confines of wetlands and small fields. This awareness came by word of mouth, from travelers in canoes, who had visited larger settlements, and in the form of occasional exotic artifacts such as lustrous copper ornaments and axes brought to the Nile from far beyond the horizon.

From the beginning, Ta-Mehu was a crossroads, its people wedded both to the Nile and to the world beyond the ocean. Perhaps the first traders to visit the delta arrived on donkeys from across the Sinai. They carried copper and semiprecious stones, shiny obsidian prized as tool-making stone, and small exotica such as lustrous beads easily carried in saddlebags. The delta and the valley upstream were treeless, except for palms, a world where everyone traveled by water in flat-bottomed boats. Prevailing northerly winds carried cargoes and people to towns and petty kingdoms up the Nile. The current brought them downstream. It was a matter of time before some of these rivercraft ventured out on the ocean, coasting laboriously to the east, where Lebanon's fabled cedars were already widely prized.

By 3500 B.C.E., some delta settlements were much larger, especially an eighteen-hectare town and cemetery at Maadi, now under a suburb of Cairo, one of several sites of this age in the region.[12] The inhabitants were above all farmers and herders, but the artifacts from the settlement reveal a kind of cultural crossroads—clay vessels from Upper Egypt and the desert, flint tools like those made in the Levant, and numerous copper objects, not only simple artifacts like needles, but also axheads that now replace stone ones, the ore coming from the southeastern corner of the Sinai Peninsula. At another center, ninety-five kilometers west of Alexandria, there are signs of an increasing social complexity, as Egyptian society developed its own distinctive beliefs and, eventually, pharaonic ideology.[13]

THE MOST DRAMATIC sea level changes after the Ice Age culminated with the final inundation of Doggerland, with the severing of the Bering Strait and the flooding of continental shelves off Asia, eastern North America, and elsewhere. Fortunately for humanity, only a few million people lived on earth, almost all of them hunters, gatherers, and fisherfolk, who yielded to the attacking sea. Except in a few more densely populated environments like Baltic shores, there was space enough for all in a warming world marked by all manner of environmental changes. By 6000 B.C.E., sea level rise was slowing, with fewer dramatic changes, except for rare events like the flooding of the Euxine Lake. Nevertheless, our vulnerability to sea level change increased slowly over thousands of years of major and minor environmental changes and steady population growth.

Nowhere does one discern this more clearly than in Ta-Mehu, the Egyptian delta, where the sea battled with the Nile. The river ponded as the Mediterranean rose. Silt accumulated, and farmers irrigated the new land. The sediment-rich soil and the eternal rhythm of the annual inundation laid the foundations for ancient Egyptian civilization. For thousands of years, the delta served as a granary, and urban and rural populations mushroomed. All was well until sea levels rose more aggressively in modern times, creating a time bomb of human vulnerability to a newly aggressive ocean.

"Marduk Laid a Reed on the Face of the Waters"

MARDUK, CREATOR OF THE UNIVERSE and of humankind, lord of thunderstorms, presided over the primordial cosmos between the Tigris and the Euphrates, the great rivers of Mesopotamia. He defeated the dragons of chaos and fashioned the spiritual and human world of the Sumerians, the first city dwellers on earth:

And then he created humans:

> *Blood I will mass and cause bones to be.*
> *I will establish a savage, 'man' shall be his name.*
> *Verily savage man I will create.*
> *He shall be charged with the service of the gods.*[1]

Mesopotamian civilization developed in a patchwork of small city-states on the floodplain between the Euphrates and Tigris in what is now southern Iraq over five thousand years ago. Sumerian lords presided over a harsh, tumultuous landscape of violent storms and sudden floods. Their civilization arose in an incubator of extreme temperatures, dense marshes, deserts, and rising sea levels, a recipe for cosmic and environmental chaos if ever there was one. Yet this complex, oft-changing landscape of conflict between salt- and freshwater nurtured some of the earliest cities on earth.

According to Sumerian legend, Marduk, "most potent and wisest of gods," mounted his storm chariot and forged order out of chaos with floods, lightning, and tempest. Order and chaos, great marshes, the threat of flood and renewed violence in the natural world and the cosmic realm: The Sumerians lived in a green and well-watered land, but at the mercy of violent rivers and changing sea levels. Only the Tigris and Euphrates with their floods made habitation possible on an arid plain besieged by climatic extremes where summer temperatures regularly soar to a humid 49 degrees Celsius (120 degrees Fahrenheit) and winters bring violent storms, heavy rain, and chilling temperatures. The great rivers emptied into a marshy delta before flowing into the Persian Gulf, the so-called Lower Sea. These marshes and the rising waters of the Gulf played a significant, even decisive, role in the rise of civilization.

THE STORY BEGINS in the Lower Sea. Twenty-one thousand years ago, at the final climax of the Ice Age, the Persian Gulf was almost entirely dry land, an arid depression with only one river flowing through its deepest axis in an incised canyon to the Gulf of Oman that was over a hundred meters lower than today.[2] Scientists call this watercourse the Ur-Schatt, made up as it was of the modern-day Euphrates, Karun, and Tigris Rivers. Where there is now a clogged river delta at the head of the Gulf there were only narrow floodplains on either side of rivers hurrying their way downstream, following relatively steep gradients.

Sea levels here rose after fifteen thousand years ago, as they did elsewhere in the world. The rise appears to have averaged about a centimeter a year until about 7000 B.C.E., when the rate of climb slowed. A centimeter a year does not seem like much, but the cumulative effects of even a quarter century's rise would have been noticeable across flatter terrain, where seawater would have spread more rapidly horizontally than vertically. Six thousand years of rapid sea level change had a dramatic effect on the basin that was to become the Persian Gulf. The first four thousand years filled the deeply incised canyon of the Ur-Schatt in its lower and middle reaches. Subsequently the rising ocean inundated the broader and shallower parts of the region. Water spread laterally at a rate of about 110 meters annually, one of the fastest such climbs anywhere in the post–Ice Age world. At times before the final slowing after about 7000 B.C.E., the lateral transgression over flatter terrain may have exceeded nearly a kilometer a year.

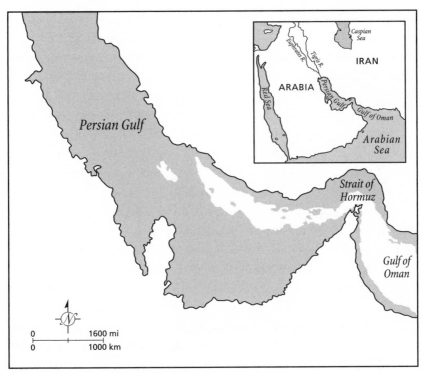

Figure 4.1 *The Persian Gulf during the Late Ice Age, ca. twenty thousand years ago.*

Rising seawater reached the present-day northern Gulf region between 7000 and 6000 B.C.E.—at about the same time as the English Channel separated Britain from the Continent and Doggerland finally disappeared under the North Sea. Shallow Gulf waters quickly inundated a low-lying delta region, turning it into a shallow marine lagoon environment, both raising the water table and shallowing the river gradients. The floodwaters of the Euphrates and Tigris still surged to the ocean, but the deep channels of earlier times now lay under ever-deeper layers of silt carried from upstream. Each summer the inundations overflowed natural levees or shallow riverbanks and flooded the flat landscape. An extensive marine estuary formed inshore of the expanding coastline. By 4000 B.C.E., the Gulf shoreline was as much as 2.5 meters higher than today in some places, so much so that geologists have identified ancient

marine sediments some four hundred kilometers north of the modern
northern Gulf coastline. One estimate places the northern Gulf coastline as far as two hundred kilometers north of its present shore as recently as 3200 B.C.E.

The sea level rise coincided with an increase in seasonal rainfall across
the Arabian Peninsula and southern Mesopotamia between about 9000
and 8000 B.C.E., associated with powerful Indian Ocean monsoon activity that lasted until about 4000 B.C.E. River runoff increased; lakes
formed between dunes on the Arabian Peninsula; pollen grains from Israel and Oman testify to less arid vegetation across the landscape. Less
direct geological clues add to the portrait of a slightly better watered
region, among them a lattice of seasonally active wadis (gullies) that
drained into the growing Gulf and the Arabian Sea. Both deep-sea
cores and terrestrial indicators tell us that humidity remained relatively
high until about 4000 B.C.E., when aridity increased gradually over
many centuries.

The changes in rainfall patterns resulted from changing monsoon
patterns brought about by a northward shift of the Intertropical Convergence Zone in the Indian Ocean.[3] Minor changes in the tilt of the
earth's axis brought warmer summers and colder winters. Around
3500 B.C.E., the monsoons weakened further and climatic conditions became much more arid. Lakes dried up; dunes formed over wide areas at
the southern edge of the Gulf. Cores from the Arabian Sea reveal major
dust storms. Greater aridity coincided with the deceleration of sea level
rise and altered the Mesopotamian environment significantly. The rivers
ponded, flowing outward when in flood, and spreading thick deposits
of fine sediment from far to the north across the flat landscape. Eventually the Euphrates and Tigris slid, as it were, to either side of the slowly
rising mound of alluvium that formed along the central axis of the valley. The groundwater table grew higher and higher every kilometer downstream. As time passed, the saline and brackish lagoon of earlier times
gradually became a huge expanse of freshwater marshes and lakes. Dense
reedbeds trapped fine silt and helped maintain the flat terrain. A complex mosaic of estuaries, rivers, wetlands, and marshes gradually formed
throughout much of what is now extreme southern Iraq. Between

3000 and 2000 B.C.E., extensive wetland areas formed along about two hundred kilometers of the rivers.

SUCH, THEN, WERE the sea level changes and climatic shifts that turned the Persian Gulf from a semiarid basin into a shallow waterway about a thousand kilometers long with a narrow entrance only thirty-nine kilometers wide in the Strait of Hormuz at the head of the Gulf of Oman in the northeastern Indian Ocean. Did hunting bands live in this challenging depression formed by low sea levels? Almost certainly there were at least narrow belts of wetlands and marshes close to the Ur-Schatt and on its floodplains by 10,000 B.C.E., which would have attracted small hunter-gatherer bands in such semiarid terrain. The river was a natural corridor from higher ground in northern Mesopotamia and the country on either side of the then-dry Gulf and in the low-lying basin itself. Around 8000 B.C.E., the Gulf was flooding rapidly as rising sea levels spread laterally from the river.[4]

The sea level rise came as human life changed dramatically with the adoption of farming and herding by people who had hunted and foraged for wild plant foods for tens of thousands of years over a broad area of the Middle East. The new economies spread rapidly as wetter conditions returned around 8000 B.C.E. Small agricultural communities appeared over wide areas of southwestern Asia between the Mediterranean and the Indian Ocean, and in the Gulf region—not that we know much about them. There are traces of small, transitory farming settlements, known mainly from stone tools, along the northeast coast of the Gulf. Tiny villages by the shores of now-dried-up lakes across the Arabian Peninsula date to between 7600 and 3500 B.C.E. By 5300 B.C.E., similar settlements also flourished at isolated spots along the Gulf's western shore.

As far as we can tell, these people herded goats and sheep and relied on sporadic farming, foraging, and small game. At strategic locations along the coast, they also collected mollusks and caught fish and sea turtles. As early as 6000 B.C.E., we know that some farmers lived in the extreme south of what is now Iraq at the edges of, and within, what were becoming some of the richest, most biologically diverse environments on earth. Since then, for more than five thousand years, humans have widened and dredged the marshes' channels, irrigated fields, and built reed houses here, atop laboriously piled-up artificial islands of bundled reeds. These marshes were to become an anchor for the towns, cities, and irrigated landscapes that came into being in later millennia.

As recently as the 1950s, there were 15,500 square kilometers of marshes at the head of the Gulf. No one knows how extensive they were in earlier times. Those who have lived on the margins over the millennia have trodden hard on the dense marshland. Sumerian rulers hunted lions here, formidable prey shot into regional vanishment only during the twentieth century. Exiles, fugitives, and rebels sought sanctuary among the reeds. Assyrian king Sennacherib captured Babylon in 703 B.C.E. and pursued its Chaldean ruler, Merodach-Baladan, southward. Sennacherib "sent my warriors into the midst of the swamps and marshes and they searched for him for five days, but his hiding place was not found." Nine years later, an expedition to Elam took Sennacherib's ships to "the swamps at the mouth of the river, where the Euphrates empties its waters into the fearful sea."[5]

Upstream of the wetlands and marshes, the landscape gave way to arid terrain, where human settlement clustered near major watercourses and on levees and natural ridges that became islands when the floods came. The scale of the inundation varied greatly from one year to the next. The nearby marshes and wetlands with their abundant plants, wildlife, and fish served as anchors for many early farming communities, such foods being the insurance against crop failure when fast-running water swept away growing wheat and barley before being absorbed by the spongelike reeds and swamps of the lower delta. It was no coincidence that numerous farming villages and the earliest cities in what is commonly called Sumer flourished within a short distance of lush marshes. This completely different, life-sustaining environment assumed near-mythic proportions in the Sumerian world as a direct result of rising sea levels.

The marshes once formed tall palisades of reeds and narrow muddy channels, interspersed with great expanses of open water, rushes, and tangled sedge. Inconspicuous defiles would burst abruptly into lakes with brilliant blue water, alive with swirling birds. I have never visited these marshes, but I know well the feeling of paddling kayaks along obscure defiles through dense reeds in other waterlogged landscapes, with only a swath of blue sky high overhead. The world feels secure, closed in, remote. The British traveler Wilfred Thesiger spent years living among the Marsh Arabs in the 1940s and 1950s when much of their traditional life still thrived. He described an environment of contrasts: "Sometimes the setting was winter, the water ice-cold under a chill wind sweeping across the Marshes from the far-off snows of Luristan. Sometimes it was summer, the air heavy with moisture, the tunnels at the bottom of the dark towering reeds where mosquitoes danced in hovering clouds, unbearably hot."[6]

The diversity of plant foods, wildlife, fish, and birds in the marshes was remarkable. At least forty mammal species, including large wild pigs and an unusual water-dependent gerbil, flourished in and around the marshes. The bird life was extraordinary. Wrote journalist Gavin

Figure 4.2 *Aerial view of the Marsh Arab village of Saigal shows individual family compounds with houses made of reeds on small islands made from packed earth and reeds in southern Iraq where the Tigris and Euphrates Rivers meet. Nik Wheeler/Flickr.com.*

Young after a visit in the 1950s: "The reeds we passed through trembles or crashed with hidden wildlife: otters, herons, coot, warblers, gaudy purple gallinule, pygmy cormorants, huge and dangerous wild pigs."[7] Forty-two bird species bred in the marshes, which were also a major wintering area for birds from western Eurasia and as far away as western Siberia. The West Siberian-Caspian-Nile flyway is one of three major flyways in western Eurasia. At least sixty-eight water bird species and fifteen of birds of prey are known to have wintered in the wetlands. The marshes of the Euphrates and Tigris were a major staging point where migrants would pause to regain their strength and build fat before moving on. After spring, most of the birds are gone.

Then there were the reeds, perhaps the most important resource of all. The giant reeds with their tasseled tops resembled bamboos, their stalks so strong that boatmen used them to pole canoes fashioned from reed bundles coated with bitumen. In a largely treeless environment, reeds were a true foundation for human existence, which is why the god Marduk used reeds to fashion the first divine and human dwelling places. Most villages in recent times were patches of artificial islands adorned with matting houses. The same must have been true of ancient settlements, for there is no other way to live in such a waterlogged environment. Families decided how big they wanted their houses to be, then gathered huge piles of reeds, which they stomped into the water, having surrounded the house site with a reed fence.[8] Once the fibrous mound emerged above the surface, the fence joined the mound, followed by yet more reeds, or lenses of mud and reeds, firmly stamped down to form a compact mass. In the end the house mounds were virtually indestructible and often remained in use for generations. When the flood came, the owners simply added more reeds to the pile and kept abreast of the rising water. The houses themselves had bent rafters of reed bundles; neatly woven mats of the same material formed the roof and walls. Anywhere near the water, people could not live without reeds on the edge of or within marshes nourished by spring floods, which could raise water levels by up to three meters. As a Sumerian chronicle remarked, "Ever the river has risen and brought us the flood." Then it adds, "All the lands were sea, then Eridu [Sumer's oldest city] was born."[9] One senses a complex trajectory of history here: rising sea levels, then ponding of rivers and sediment once swept to the sea settling and forming not only desert but a marshy floodplain. And as part of the trajectory, farmers settled at the edges and within the marshes. Their remote descendants helped create a literate urban civilization.

To SAY THAT rising sea levels created a civilization or that the Marsh Arabs of the twentieth century and their homeland are surviving mirrors of people and places of five thousand years ago is, of course, a woefully inadequate explanation of very complex events. Nevertheless, there can be no question that the expansion of the Persian Gulf played a decisive role in the first human settlement of what became Sumer. For generations, we archaeologists have argued that the first farmers colonized the delta from the north, bringing irrigation agriculture with them, perhaps around 6000 B.C.E. Perhaps, however, one can argue this a different way, with farmers following wetlands northward from the drowning Gulf, then settling on its margins near the great rivers or on humanly constructed mounds within the marshes once sea levels stabilized. Such a scenario is little more than intelligent guesswork. Excavation and survey within the waterlogged landscape is near impossible, and in any case, most farming villages of the day lie deep below river alluvium. Only occasionally do we get a glimpse of what life must have been like when sea levels were at their height and marshes tended to define life in the south. Almost invariably, these brief portraits come from the bases of long-occupied, ancient city mounds—some of the places where the gods were said to have cast reeds upon the ground and formed habitations.

Three quarters of a century ago, a charismatic British excavator, Leonard Woolley, spent twelve seasons from 1922 to 1934 excavating the ruins of the ancient city of Ur, biblical Calah, associated in the Old Testament with Abraham. Woolley excavated on a grand scale.[10] Hundreds of workers cleared overburden accompanied by lilting songs chanted by a Euphrates boatman recommended by his legendary foreman Sheikh Hamoudi, whose passions in life were said to be archaeology and violence. A decisive leader who was never at a loss, Woolley

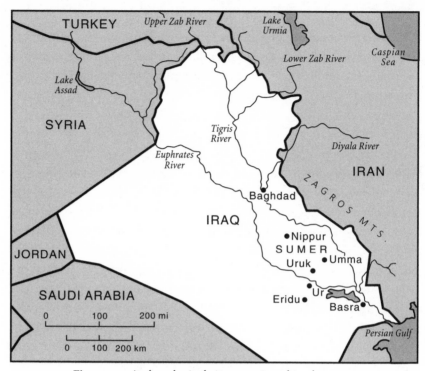

Figure 4.3 *Archaeological sites mentioned in chapter 4.*

worked with only a handful of European colleagues, moving earth by the ton. He excavated the great ziggurat pyramid of Sumerian king Ur-Nammu, built during the twenty-first century B.C.E. and still towering above the ruined city. His men cleared entire city quarters, where Woolley delighted in showing visitors the cuneiform inscriptions set in the doorsills that identified their owners.

This remarkable archaeologist became world famous for his excavation of the spectacular royal graves under the city, notably that of Prince Puabi. Woolley was a vivid writer, who brought the past to life with effortless panache—even if some of the details did not entirely match his observations. He described the solemn royal funerary processions, the ordered courtiers being handed clay cups and simultaneously taking poison to accompany their lord into the afterlife. Spectacular stuff by any archaeological standard, unfortunately much of it unverified, for Woolley's excavation notes are too incomplete by modern standards. But the royal graves momentarily paled into insignificance alongside the discoveries at the bottom of a deep pit sunk to the very base of the great city mound.

As part of his study of the chronology of Ur, Woolley had excavated a small mound called al-'Ubaid about six kilometers north of the city. Here a previous excavator had found a very early temple built by Ur's first Sumerian king that clearly merited further excavation. The new excavations probed a low tumulus about 1.8 meters deep some fifty-five meters from the temple. A small village—built on a low hill of clean river silt that stood above the floodplain—emerged from the mound, marked by hundreds of painted potsherds, the remains of small matting houses, and tools made of obsidian (volcanic glass), but no metal artifacts whatsoever. In the days before radiocarbon dating, Woolley could not date the village, but it was earlier than the Sumerian temple and anything he had so far discovered at Ur itself. He assumed that these humble farmers were the first inhabitants of southern Mesopotamia long before the first cities rose along the Euphrates.

After excavating al-'Ubaid and the royal graves, Woolley turned his attention to the base of the Ur mound. A small test trench yielded 'Ubaid-like potsherds on sterile sand underlying deep flood deposits. The thick zone of mud puzzled Woolley. His staff was also nonplussed. Then his wife, Katherine, came by. She glanced at the trench and remarked casually, "Well, of course, it's the Flood." In an era still somewhat obsessed with proving the historical veracity of the scriptures, Woolley had wondered the same thing, but "one could hardly argue for the Deluge on the strength of a pit a yard square."[11]

He returned in 1930 and sunk an enormous pit that went down nearly twenty meters to bedrock. The great trench yielded a stratified chronicle of ever-earlier settlement that had culminated in the early Sumerian city. First the diggers uncovered eight levels of houses, then the thick layers of a potting factory where the potters had discarded their "seconds" and constructed new kilns atop the broken vessels. Woolley

Figure 4.4 *The so-called Flood Pit at Ur, excavated by Leonard Woolley. © University of Pennsylvania Museum.*

tracked the changes in pottery styles as the diggers went deeper. He started with the wares found in the houses, then the vessels from the potters' precinct—plain red pots and vessels fashioned from greenish clay with red-and-black-painted designs, just like those from nearby al-'Ubaid.

Some of the owners of the red vessels lay in graves dug into sterile river silt under the potters' workshop. The dig passed through nearly three meters of clean river silt deposited by the Euphrates. Below lay three superimposed levels of mud-brick and reed houses identical to those at the original al-'Ubaid site. At last the excavators reached stiff green clay, once a swampy marsh—what Woolley called "the bottom of Mesopotamia." Apparently the first occupants of Ur tipped their garbage into the marsh, built up a low mound, and then settled on it, inadvertently forming the core of a much later city.

The excavation of what soon became known as the Flood Pit at Ur gave Leonard Woolley a unique chance to exercise his fluent pen. With effortless imagination, he pulled out all his literary organ stops. Here, he claimed, was not only evidence of a great flood then popularly associated with the Garden of Eden but also archaeological confirmation of the great inundation described in Assyrian and Sumerian epics preserved on clay tablets. The 'Ubaid people of preflood times had not been obliterated by the biblical deluge, but had survived the waters to plant the seeds of Sumerian civilization. "And among the things they handed down to their successors was the story of the Flood . . . for none but they could have been responsible for it."[12] Even Sumerian king lists spoke of rulers who reigned before and after what must have been an epochal flood, even if it was not the one in the scriptures.

The silt deposit was deepest on the north side of the city, where the mound had broken the force of the inundation. He estimated that the flood covered an area at least 480 kilometers long and 160 kilometers across, destroying hundreds of villages and small towns. Only the oldest cities were safe on their ancient mounds. The disaster must have been catastrophic and endured in human memory for generations. "No wonder that they saw in this disaster the gods' punishment of a sinful generation and described it as such in a religious poem."[13]

The Flood Pit with its evidence for a huge inundation caused an international sensation at the time, overshadowing similar flood deposits found elsewhere in the depths of other city mounds. Today both archaeologists and linguists doubt that the Ur floods, or any others for that matter, were the source of Mesopotamian flood narratives. They think they are merely evidence for endemic, sometimes catastrophic inundations in the flat terrain of southern Mesopotamia, where the gradient of the floodplain changes by a mere thirty meters over seven hundred kilometers.

Today, prosaic finds like stone artifacts and characteristic greenish, painted 'Ubaid pottery like that from Ur document a far-flung, simple farming culture that flourished over a wide area of southern Mesopotamia. Upstream of the wetlands and marshes, the landscape gave way to arid terrain, where human settlement clustered near major watercourses and on levees and natural ridges that became islands when the floods came. Nearby marshes and wetlands with their abundant plants, wildlife, and fish served as anchors for many early farming communities, such foods being the fallback when fast-running water swept away growing wheat and barley before being absorbed by the spongelike reeds and swamps of the lower delta.

This insurance worked well, for within two thousand years or so profound changes were under way in southern Mesopotamian life. The Gulf was retreating slightly; life-sustaining marshes with their vital food staples may have proliferated as freshwater wetlands replaced brackish lagoons. From the earliest days of farming, people living by the Euphrates and Tigris Rivers had diverted floodwaters through narrow furrows onto their small fields, lying strategically close to side channels. Barley and wheat grew well in the fertile soils and the floods of early summer came at a critical moment for the growing crops. As the centuries passed, small clusters of reed dwellings became larger, more permanent communities built on levees and low ridges. Hamlets coalesced, then turned into larger villages and small towns, clustered around shrines to patron deities, just like the humble settlement on the low mound at al-'Ubaid near Ur. No one living in these growing communities had any illusions about the risks of farming in an environment where the capricious rivers nearby could flood and wipe out generations of close-set houses in a few hours, then change course without warning.

A combination of irrigation and food from marshes worked so well that some of the larger towns became the world's first cities. Uruk was the home of the mythic Gilgamesh, hero of the epic that bears his name, and one of the earliest cities on earth. The city's weathered mounds lie in an arid landscape east of the present-day Euphrates, on the banks of the now-dry but ancient Nil Channel, which provided both irrigation water and access to trade routes up- and downstream. Canals separated Uruk's bustling neighborhoods, so much so that some imaginative modern observers have called Uruk the Venice of Mesopotamia. According to the *Epic of Gilgamesh*, Uruk had three segments, "one league city, one league palm gardens, one league lowlands, the open area(?) of the Ishtar Temple."[14]

The city originally came into being when two large villages merged into a single settlement around 5000 B.C.E. Serious growth began a thousand years later, when crowded neighborhoods of small houses clustered around major shrines. By 3500 B.C.E., Uruk was much more than a large town. Satellite villages, each with their own irrigation system, extended as much as ten kilometers in all directions. Uruk and its temples had become a major trading center, linked with communities near and far by trade that moved up and down the great rivers. However, throughout its life, the thriving city relied on the nearby marshes, which figured prominently in incantations and myths, for it was from the reeds that the god Marduk had created the first dwellings in this violent land:

Marduk laid a reed on the face of the waters,
He formed dust and poured it out beside the reed.
That he might cause the gods to dwell in the habitation of their hearts'
*   desire.*[15]

These reeds would never have grown had it not been for the rise and fall of the Lower Sea to the south.

Distant, unseen forces governed the fate of Sumerian cities. For thousands of years, rising sea levels downstream rendered much of the flat landscape south of Ur unsuitable for human habitation, except where natural ridges lay above flood level. Then the Gulf shoreline retreated significantly after 4000 B.C.E. Marshes formed and became an anchor for human life, while the rivers remained an unpredictable force. Both the Euphrates and the Tigris meandered over the landscape in broad loops and sometimes changed course when they burst their banks during major floods. The Euphrates has shifted its lower reaches time and time again without warning, transforming both natural and humanly altered landscape in fundamental—sometimes catastrophic—ways. Each village and every city faced a daunting landscape that could change in days. No river channels in southern Mesopotamia were permanent. One rapid shift saw the Euphrates redirect its course abruptly from east of the growing city of Nippur to much farther west, leaving the city literally high and dry. Uruk suffered greatly when the Euphrates shifted eastward from a line through the city to a more easterly course that favored another urban center, Umma.

All of this made for a volatile political and social environment and for haphazard settlement patterns that followed elongated levees along major channels and minor ones that ran parallel to the main channel. Mesopotamia was a volatile, ever-changing landscape, whose configuration changed with rises and falls in Gulf sea level and with the severity of river flooding. This helps explain why Sumerian civilization was a jigsaw of intensely competitive city-states, each with its own hinterland of villages and irrigated lands, and completely dependent on the vagaries of the natural environment and the ocean that lay far over the horizon.

Here, rising sea levels helped foster farming and civilization, but at the same time created environmental problems—among them rising salinity—that have plagued farmers between the Euphrates and Tigris Rivers ever since. Here, too, climate change, in the form of shifting rivers and rising sea levels, was a crucible for intercity strife and petty wars. For the first time in history, standing armies became instruments of rulers' policy. Here and elsewhere, warfare was to become endemic in human life. For the first time, too, shifting patterns of international trade made anchorages and ports at the ocean's edge major players in the rise and fall of civilizations, and in human relationships with the ocean.

Catastrophic Forces

Around six thousand years ago, the world's oceans had reached more or less modern levels. Dramatic geographical transformations were a phenomenon of the past, except for local crustal adjustments, lengths of coastline elevated or lowered by earth movements, and natural subsidence. The sea still attacked, but in different ways in human terms. Those who dwelled by exposed coasts had always experienced violent storms, storm surges, and tsunamis, but their numbers were small, most communities little more than long-occupied fishing camps, farming villages, or small towns. There were casualties from surges and tsunamis, from natural catastrophes wrought by the ocean; there always had been and always will be. Nevertheless, the numbers of coast dwellers and people living close to sea level were minuscule even by the standards of two thousand years ago when Rome ruled the Mediterranean world and Emperor Qin Shi Huangdi unified China into a single kingdom. The casualties wrought by severe weather events and natural cataclysms rose dramatically with the founding of the first cities some five thousand years ago, as well as with the rising importance of long-distance commerce, much of it carried by ships sailing from ports large and small.

The next six chapters examine some of the changes wrought in human societies by extreme events that originated in the ocean after six thousand years ago. Our chronological starting point is around 5000 B.C.E., usually later. For the first time I venture at times from the past into the present, something we dwell on at greater length from chapter 11 onward. Where I move from past to present is where I perceive a degree of continuity in the story that is worth pursuing in, say, the wider context of Mediterranean history or in China, where the Yangtze River shaped the birth of modern-day Shanghai. Here too we begin to define some of the issues that confront nations living close to sea level. Do you yield to the attacking ocean, staying where you are and adapt, or wall yourself off from rising sea levels and violent storm surges? These issues have been at the forefront of Venetians' minds for many centuries, have affected government policy and sea defenses in the Low Countries since medieval times, and have resonated through the halls of the British Raj in India. The chapters that follow are, as it were, the starter for the main course of today's inundations, the closing chapters of the story.

As before, we begin our journey in northern Europe.

"Men Were Swept Away by Waves"

DOGGERLAND FINALLY VANISHED in about 5500 B.C.E. as the last remnants of the Dogger Hills disappeared beneath the heaving waters of the rising North Sea. This was not a dramatic event, just a gradual encroachment of saltwater on low-lying terrain. As groundwater and sea levels rose, so inland swamps and freshwater lakes formed along the coasts of what are now the Low Countries. A layer of peat developed over many centuries, but never became very thick. The rising North Sea also turned much of the coastal landscape into muddy tidal flats and extensive tracts of landscape that dried out at low tide. When sea level rise slowed, the surf threw up sandbars that eventually became dunes. Eventually lagoons close to the coast silted up and became freshwater marshes where peat continued to grow, turning much of the Netherlands into a huge bog. Today two thirds of that densely populated modern country is basically an alluvial plain that is vulnerable to flooding.

The rising sea encroached inexorably on the land, but human existence continued much as before. As they had in Doggerland, small hunting bands frequented lakes, marshes, and wetlands where plants, fish, game, and fowl could be found at all seasons of the year. Many places along the low-lying coastlines of both Britain and the Low Countries supported a great diversity of productive environments during the five millennia between the disappearance of Doggerland and the Roman colonization of Europe. Along what is now the Belgian coast, a patchwork of tidal flats, lagoons, and freshwater peatlands lay behind a belt of sand dunes. Farther north, in the western Netherlands, four major river estuaries, each with extensive salt marshes and mudflats, broke up a virtually continuous sand dune barrier. Freshwater peat bogs developed in the sheltered conditions behind the natural blockade. A more open coastal plain, protected in part by low offshore islands, marked the northern Netherland and German coasts. Here again, mudflats and salt marshes dominated the landscape, interspersed with freshwater swamps. Eastern England's rivers and swampy estuaries also supported hunting groups for thousands of years. In these lowland landscapes, fish runs, migrating waterfowl, deer hunting, and the seasons of plant foods, not the ocean, set the pattern of human existence for nearly a thousand years.

THE NETHERLANDS COAST, midwinter, 3800 B.C.E. A bitter wind howls across the salt marsh, carrying the sound of a roaring sea on a fast-rising tide. Snow flurries cascade with the storm; the gusts strengthen and turn swirls of snowflakes into whirlwinds. A narrow levee overlooks the turbulent ocean raging across the flat landscape, a slender bastion set against the gale. A huddle of grass-and-reed huts crouches atop the ridge, set in a circle within a wooden enclosure crowded with cattle and sheep. As the hours pass and the tide rises, the encroaching tide attacks the levee. Great waves break against the low ridge; spray breaks high over the low-set dwellings; waterlogged cattle low in distress. A few breakers sweep into the margins of the settlement and flood one of the huts, which collapses in a confusion of reeds and sticks. Drenched to the skin, the family inside quickly make their way to another shelter on the far side. The villagers anxiously watch the advancing surge, knowing that eventually the tide will turn and the sea recede.

IT'S CONJECTURE, OF COURSE, but a reflection of a new reality: Farming introduced a completely new dynamic to life along the North Sea. Hunting bands could ebb and flow with the coastline, their possessions readily portable, their lives attuned to mobility. But farming anchored people to the land, to their growing crops, to the pastures where their flocks and herds grazed. Now the North Sea became far more than background noise to daily existence. The ocean became a potential enemy that could wipe out fields and flood pastures in hours, leaving people hungry.

At some point, perhaps as early as 4700 B.C.E., a few hunting groups acquired cattle and perhaps sheep from farming communities, presumably on the uplands.[1] What compelled them to do so remains a mystery. Perhaps rising sea levels and growing populations created pressure on hunting territories and led to food shortages. Thus, it would have been a logical step to broaden the food supply by herding domesticated animals. Presumably only very small numbers of animals were involved at first, perhaps acquired from farmers on higher ground, who drove their flocks and herds onto salt marshes during summer. Several centuries passed before cereal crops came into use in the wetlands. Judging from excavations at Swifterbant near the Zuiderzee, the people cleared vegetation on clay levees, then grew summer crops of emmer wheat and barley in small fields. They also harvested large numbers of wild plants, including hazelnuts, which may have been an important staple, provided one could store them. A cubic meter of hazelnuts is sufficient to provide 10 percent of the annual energy needs of a mixed population of twenty people.[2]

At first the changeover made little difference to seasonal routines. Families moved their cattle and sheep to salt marshes and flatland grazing in spring. Come winter, each community must have driven its flocks and herds as far above high tide as possible. As both cereal farming and herding took hold, so the human grip on wetlands became less transitory. When sea levels receded slightly, small groups of people would settle permanently on the better-drained fringes of peat bogs and on raised creek levees. Such locations enabled them not only to eke a living from their animals and fields, but also to exploit wild plants and nut harvests, waterfowl and fish of all kinds if they wished. By the first millennium B.C.E., permanent settlement was commonplace, even if human populations were still thin on the ground. There was good reason. Even modest but lasting farming settlement near the coast brought a quantum jump in vulnerability to major storms, extremely high spring

Figure 5.1 *Map of locations mentioned in chapters 5 and 14.*
The islands around Strand, Germany, are not included.

tides, and fast-moving sea surges that raged far inland carrying everything before them. Life became a permanent tussle against the North Sea, a war that continues to this day.

Backbreaking labor and constant vigilance were the only human weapons against the ocean. Building even modest dikes consumed days of brutally hard work in landscapes where the only natural defenses were sand dunes and elevated levees. The problem was particularly acute along the northern coasts of the Low Countries, where tidal flooding and sea surges were endemic. Such events might lay waste everything before them, but they did lead to the formation of elevated coastal marshes with freshwater bogs behind them. As the drainage improved, so farming villages appeared along the Dutch coast by the sixth century B.C.E. and along the northwestern German shore a few centuries later. Each settlement lay on elevated marshland. As water levels rose slightly after 1000 B.C.E., some farmers left home, but others fought back with artificial raised mounds made of turves and clay known as *terpen* (singular, *terp*), which translates as "staying above water," located in places where they could escape high tides most of the time.[3] The number of such mounds rose and fell as water levels fluctuated, to the point that by the fourth century C.E., a time of rising shorelines, only the most elevated coastal marshes were still occupied.

Many terpen had long histories. At Ezinge near Groningen, the first settlement of the fifth century B.C.E. lay on the surface of a salt marsh, set within a palisade about thirty meters across.[4] Three rectangular buildings with wattle-and-daub walls and thatched roofs lay inside, along with a barn set on wooden piles to keep stored crops dry. The next settlement some three centuries later stood on a mound of turf sods about 1.2 meters high and 35 meters wide. Four farms huddled together on the terp for protection against winter storms. Later farmers repeatedly added to the mound, which eventually reached a height of 2.2 meters and a width of 100 meters. During Roman times, the tumulus grew even larger, but was still a huddled group of four farmsteads that perished in a fire during the third century C.E.

Most terpen lay close to the seaward side of the salt marshes, often on low ridges running parallel to the shore. No less a literary personage than the Roman eminence Pliny the Elder wrote in his *Natural History* of such settlements, located in the "regions of the far north . . . invaded twice each day and night by the overflowing waves of the ocean . . . Here a wretched race is to be found, inhabiting either the more elevated spots of land or else eminences artificially constructed and of a height to which they know by experience that the highest tides will never reach." The mound dwellers had no flocks and wove nets from sedges and rushes "employed in the capture of the fish."[5]

Figure 5.2 *Artist's reconstruction of a terp at Ezinge, the Netherlands, ca. two thousand years ago.* © *Bob Brobbel.*

Archaeology tells us that Pliny was wrong in calling the terpen dwellers mere fisherfolk. The mound people were expert farmers, growing barley and other crops such as flax that can tolerate quite salty soils. They maintained herds of cattle and also sheep, which thrived in brackish marshlands. Wetland agriculture was a world unto its own, practiced by people who probably kept to themselves with little contact with outsiders.

The Romans found little attractive in the Netherlands' flood-beset swamps and dank woodlands. There were no minerals; the agricultural potential appeared low, and the population sparse. Only its geographical position was important, for the mouths of the Maas and the Rhine Rivers led deep into western Europe and hostile Germanic tribes could travel along them into the rich landscapes of Gaul. Julius Caesar was the first to attempt a stabilization of the northern frontier in 57 B.C.E., but the Roman hold on the Netherlands was never much more than nominal. Frequent campaigns against Germanic tribes sometimes ended in disaster, among them a campaign in 12 C.E. A Roman general, Publius Vitellius, set off for winter quarters through marshy terrain during a severe North Sea gale. The historian Tacitus tells us how:

> after a while, through the force of the north wind and the equinoctial season, when the sea swells to its highest, his army was driven and tossed hither and thither. The country too was flooded; sea, shore

fields presented one aspect, nor could the treacherous quicksands be distinguished from solid ground or shallows from deep water. Men were swept away by the waves or sucked under by eddies; beasts of burden, baggage, lifeless bodies floated about and blocked their way.[6]

MANY PEOPLE ASSUME that the North Sea and the Baltic stabilized after the flooding of Doggerland and remained largely unchanged since about 5000 B.C.E. This assumption has led to hysterical predictions of disasters caused by the seemingly unique (but real) sea level rises of today attributed to humanly caused global warming. In fact, northern coastlines have changed irregularly and unpredictably over the past seven thousand years. We will probably never be able to chronicle all the minor sea level changes that have affected the North Sea, except in the most general terms. Local topography, the configuration of the shore, tidal streams, and adjustments in the earth's crust all play significant, and still little understood, roles in the process. So the brief summary of what happened after 1000 B.C.E. that follows is cursory at best.[7]

During the late first millennium B.C.E., salt marshes and mudflats formed most of the coastal wetlands in northwest Europe, with extensive freshwater swamps behind them in many areas. By the first century B.C.E., there was a period of relatively stable, even falling, sea levels throughout eastern Britain and in much of the Low Countries. However, during late and post-Roman times, there are widespread signs of sea level rises between the third and fifth centuries C.E., which resulted in the inundation of numerous marshland settlements and once-settled agricultural land. Whether the abandonment of these villages was due entirely to sea level rise is questionable, as this was also a period of major economic and social upheaval. Over many centuries, people living close to sea level either reconciled themselves to settling on higher ground or counterattacked the sea with the first humanly fashioned defenses.

At first, farmers on both sides of the North Sea had settled on locally slightly higher terrain close to natural canals, creeks, and ditches. By judicious, communal effort, the villagers could drain water from small tracts of land, which were mainly used for grazing cattle. They could also construct low embankments to protect areas from summer floods. The terpen of the northern Netherlands and Germany were a logical extension of these landscape modifications, whereby farming villages lay permanently within tidal landscapes. From there, it was a simple step from marshland modification to the partial transformation of landscape by constructing earthen seawalls. Once a low earthwork was in place, the farmers could then gradually enclose land on a piecemeal basis as the local population rose. Alternatively, and more rarely, they could reclaim large areas of land as a planned, systematic enterprise.

At first there were no seawalls along the Netherlands coast. Nevertheless, farmers attempted to control flooding with dams and sluices. At Vlaardingen in the western Netherlands, a first-century landscape of paddocks and farming settlements lay along tidal creeks.[8] Dams constructed of clay sod with layers of reeds and sedges and revetted with sharpened stakes blocked creeks and side drainage channels. Hollowed tree trunks allowed freshwater to flow under the dam, each with a hinged wooden flap valve at the end of the culvert.

Land reclamation, even on a small scale, required long experience and an intimate knowledge of tidal streams, wave patterns, and natural drainage. Even a modest earthen seawall a few meters high was an ambitious enterprise, at best a high-risk endeavor. Months of backbreaking digging through layers of gravel and piling up thick clay and turves in small basketloads in all weathers could be swept away in hours. Just planning the job could take days of argument and discussion. Who would undertake foundation digging, the collection of turves, even the weaving of strong baskets for moving soil? Above all, what were the long-term benefits for the community, and perhaps its neighbors? We can imagine long arguments and counterarguments among people who had no illusions about what lay ahead, not only in constructing, but also in maintaining the low dikes once they were in place. Everyone knew that the long-term rewards could be significant, even for generations as yet unborn, but the risks were significant. Seawalls were vulnerable to high tides, to sea surges, and they placed a heavy burden of watchful maintenance on those who lived behind them.

Centuries of hard-won experience came into play during seawall

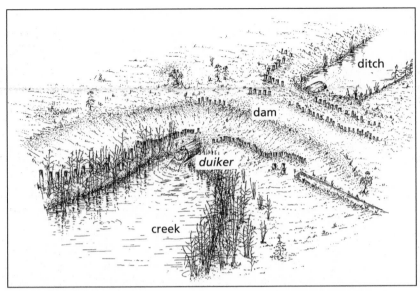

Figure 5.3 *Artist's reconstruction of a Roman period dam and drain that crossed a tidal creek at Vlaardingen, Netherlands, ca. 75–125 C.E. The drain, or duiker, consisted of two hollowed tree trunks, one fitting into another. The hinged flap was at the outer end.* © B. Koster/Vlaardings Archeologisch Kantoor.

construction. Each formed a continuous barrier that prevented inundation year-round and reduced waterlogged acreage. The absence of water improved soil aeration and helped warm the earth more rapidly in spring. As a first step, major flooding sources had to be controlled, not only tidal flow, but also water runoff from higher ground. Once this level of control was in place, constructing drainage systems could lower water tables. All of this was a laborious, slow-moving task carried out for the most part by individual communities. Fortunately for the diggers, early seawalls along the North Sea were relatively modest because of lower sea levels at the time. Exact dimensions are hard to come by owing to later sea surges and tidal action, but even as late as the eleventh and twelfth centuries C.E., an embankment near Enkhuizen in the northern Netherlands was only 1.6 meters high compared with the 6 meters of today.[9]

To build an earthwork like a dike or seawall required digging the raw material from nearby marshes and the foreshore. Each community felled hundreds of trees inland for bridges and sluices, then dragged the trunks along narrow tracks through marshy terrain. Much remained to be done when the seawall was complete. While the builders could make some use of natural drainage channels behind the embankments, they also had to create an entirely artificial system. Even when the seawall was finished, the work never ceased. The defenses had to be inspected regularly, especially after winter storms. Local villagers had to clean and scour the drainage ditches regularly by hand. Fortunately, the sludge could be spread on nearby fields to improve soil fertility. The risk of sea surges and catastrophic flooding was always present.

Early reclamation efforts usually involved simple earthen embankments built of marsh clays, usually strengthened with timber stakes, wattle fences, stones, and straw. The work had to be completed rapidly; winter storms soon washed away unfinished fortifications. Once the seawall was complete, water flowing from higher ground had to be channeled across the reclaimed land, either by digging a channel or, in some cases, by using a raised watercourse. Seawater drained through elaborate systems of gullies and ditches, often using a combination of plowed furrows and larger drainage ditches excavated by hand that also served as field boundaries. The same defiles provided water for livestock. Within a few years, all the salt was washed from the reclaimed soil, which could then be used for pasture or enriched with lime, marl, or manure to turn it into arable land. All this effort could come to naught in the face of a spring tide, a winter gale, or a sea surge that swept away dikes, dwellings, flocks, and people.

The threat from the sea was always there. As Christianity took hold, so natural disasters became a symbol of the wrath of God. No one could predict great gales and storm surges. High waves and coursing tides could return year after year or remain just a silent menace in the background for generations. Most such disasters have long vanished into historical oblivion, but a few memorable cataclysms survive in medieval records. In 1014, *The Anglo-Saxon Chronicle* describes how in eastern England "in that year, on the eve of St. Michael's mass, came the greatest sea flood wide throughout this land, and ran so far up as it had never before had done, and washed away many towns, and countless numbers of people."[10]

BETWEEN 500 AND 800 C.E., falling sea levels led to the formation of the so-called Young Dunes, which now form a practically unbroken coastal rampart from Den Helder in the north to the Hook of Holland in the south. Then an unexpected sea level rise penetrated deep into dune country between 800 and 950 C.E., at which point serious dike building finally began. Because of the higher sea levels, terpen in the northern Netherlands and northwestern Germany were at their most numerous during the tenth century, many of them built on marsh bars and alluvial ridges. Some now reached a large size, surrounded by a circular road and ditch. A freshwater pond usually lay in the middle of the village; wells lined with wooden barrels sunk deep into underlying peat and sand provided additional water supplies. The villages looked like islands, surrounded by orchards and trees, dominated by a church tower, in the midst of treeless salt pastures.[11]

Around 1000 C.E., the North Sea again receded slightly, giving some respite from storm surges and warfare. Populations rose. Many communities now began the task of protecting their lands from both storms and the insidious penetration of saltwater inland. The first dikes were little more than raised trackways that led from one terp to another, soon extended to form closed systems of water defenses. Simple barriers crossed creeks or ditches, at first fully removable and then replaced by simple wooden sluices that opened and shut with ebb and flood.

Despite these sea defenses, repeated flooding attacked the region during the sea level rise of the slightly warmer late Medieval Warm Period after 1000 C.E. In the north, tens of thousands of people lost their lives during a storm surge between Stavoren, now on the Zuiderzee, and the mouth of the Ems River in 1287. Subsequent inundations flooded huge tracts of coastal land. The only defense at the time was to raise terpen heights. Ezinge was now 18 meters above the surrounding landscape and nearly 425 meters across.

To the south, the waning sea level rise between 950 and 1130 coincided with a rise in population at a time when the threat of constant Norse raids receded. Farmers needed more cultivated land; drier, less severe winters and a decrease in storminess led to aggressive settlement along bogs by burning off vegetation, then digging ditches for drainage and property boundaries. Extensive colonization transformed hitherto un-inhabited peat swamps into farmland. Soon major landowners became increasingly involved in larger-scale land reclamation.

AROUND 1130, AN increase in storminess and the resumption of sea level rise caused havoc. As the unpredictable sea surges continued, indi-vidual communities cowered behind their own, usually inadequate, de-fenses. Perhaps they felt powerless in the face of divine wrath, which gave them little incentive to organize their water defenses. Few villagers looked farther afield than their own fields and grazing grounds, or thought of anything but their own parochial interests. Prominent land-owners and religious houses led the way and did most of the larger-scale diking. Thanks to their efforts, sea defenses slowly extended along lengthy stretches of the coast and blocked river breaches, many of them haphazardly.

Building seawalls and dikes was one thing; maintaining them was an-other. Lethargic farming communities neglected their own water defenses, perhaps in a fatalistic belief that they were doomed to suffer God's anger. At the back of every Christian's mind lurked the terrible image of the Last Judgment, "when the revelation of that day will be very terrible to all created things."[12] Much of the time, instead of banding together, farmers and landowners wrangled constantly over the cost of maintain-ing dikes and did nothing. Inevitably, neglected sea defenses fell into disrepair; inevitably, too, unpredictable floods destroyed whole villages and killed entire families.

Unbeknownst to the quarreling farmers and landowners, who knew nothing of the wider world, the great sea surges originated with the strong winds generated by extratropical cyclones far out in the Atlantic. Such tempests covered an enormous area; gale-force winds blew over hundreds of kilometers of ocean. The surges that beset them and still beset the North Sea typically originated in storms off northern Britain. They moved southward along the eastern shores of Scotland and England and up the Thames River. Intense low pressure and strong northerly winds drove the North Sea's surface waters southward into the narrowing sea basin, piling up water in vicious assaults against the Low Countries.[13] The most destructive storm surges coincided with unusually high spring tides, when incoming seawater flowed far inland, checking the discharge of freshwater from higher ground. Violent waves and scouring tides would sweep ashore carrying everything before them, causing human tragedies that would assume almost mythic status in folk memory.

We know of major floods from at least three violent storm surges that hit the German and Dutch coasts in about 1200, 1219, and 1287.[14] The surge of January 16, 1219, the feast day of St. Marcellus, killed at least thirty-six thousand people. By bizarre coincidence, one of the greatest and best known medieval surges, known as the Grote Mandrenke (the Great Killing of Men) of 1362, struck on the same day as the 1219 cataclysm:

"Then on the morrow of Saint Martin . . . there burst forth suddenly at night extraordinary inundations of the sea, and a very strong wind was heard at the same time as unusually great waves of the sea. Especially in places by the sea, the wind tore up anchors and deprived ports of their fleets, drowned a multitude of men, wiped out flocks of sheep and herds of cows . . . an infinity of people perished, so that in one not particularly populous township in one day a hundred bodies were given over to a grievous tomb."[15]

The Grote Mandrenke originated as a severe southwesterly gale in the Atlantic. Storm winds swept across Ireland, then England, where houses collapsed and the wooden spire of Norwich cathedral fell through the roof below. Across the North Sea, the gale descended on the Low Countries and northern Germany at high tide. Wind and waves wiped out sixty parishes in coastal Denmark and widened the entrance of the Zuiderzee to the ocean. This was the only positive outcome of the disaster. Once stabilized by dikes, the new inlet became a major hub of maritime trade for the later Dutch Empire. A huge sea surge swamped harbors and landing places without warning. The wealthy port of Rungholt on the island of Strand in Schleswig-Holstein, northern Germany, vanished below the waves. Over half of Strand's population drowned.[16] On the opposite side of the North Sea, a trading village, Ravenser Odd at the mouth of the Humber River on the Yorkshire coast, also disappeared. The inevitable stories of ringing church bells below the waves heard by passing sailors persisted, as if the sea surge resulted from divine wrath. A minimum of twenty-five thousand people, certainly many more, perished in the Great Drowning.

Storm surges were even more frequent between the thirteenth and fifteenth centuries, such as the apocalyptic St. Elizabeth's flood of November 1421. Persistently stronger winds moved coastal dunes several kilometers inshore from Texel in the north to Walcheren to the south. Only since the fifteenth century has the building of large breakwaters on the foreshore and the planting of grasses and trees on the dunes slowed the process. The destruction was cumulative but immense. For instance, the church at Egmond, an important agricultural settlement, lay in the lee of the dunes in 1570. The dunes moved beyond the church, leaving it exposed, until it fell victim to the North Sea in 1741.

By the end of the Middle Ages, intricate rings and lines of dikes, also river embankments and dams, formed a measure of coastal protection, but not nearly enough for a region of growing cities and rising population. A confusing mélange of dike, drainage, and supervisory boards oversaw the work, which fostered powerful notions of independence and self-government, and also of service for the common good, quite independent of feudal lords—something radically new. Major landowners now offered incentives such as low rent or free status to anyone willing to settle in such flood-prone areas.

Land reclamation had also lowered the level of peat surfaces over hundreds of square kilometers. The authorities outlawed turf cutting as part of peat digging for salt extraction within a few kilometers of dikes in 1404, on the grounds that it seriously undermined sea defenses.[17] Many rivers now flowed above adjacent lands, making it easier for floodwaters to breach the dikes. Rising water tables had turned much agricultural land into waterlogged pasture or hay meadows. With only gravity to rely on, draining the inundated land often proved near impossible. Land reclamation now came into play on a much larger scale.

The Grote Waard, south of the modern city of Dordrecht, became the most highly developed reclamation project.[18] Once a swampy waste, with small villages on higher ridges linked by dikes that served as tracks, Dirk III, Count of Holland, seized the land during the eleventh century. By 1300, his successors had dammed the main streams and built a ring of dikes. A century later, over forty villages sent grain, turf, and salt to Dordrecht's markets from a single great polder covering some fifty thousand hectares.[19] (A polder is a tract of reclaimed land protected by a dike.) But the Grote Waard carried the seeds of its own destruction. The poorly maintained, often undermined dikes built on unconsolidated peat collapsed on themselves if built up too high. Much of the reclaimed land lay at a vulnerable spot, where floods could attack it on two sides, both from the North Sea and from the waters of great rivers like the Maas, Rhine, and Waal. Years of major surges had breached the dikes again and again throughout the fourteenth century, but the farmers repaired the damage and drained the land anew. Then a major inundation in 1420 destroyed all the records of the water board and disrupted its administration at a time of major political strife.

A year later, on the night of St. Elizabeth's Day, November 18–19, 1421, fierce westerly winds coincided with an exceptionally high tide, breached the dikes, and inundated the Grote Waard when the Maas and Waal Rivers were at high flood levels. The rising seawater burst dikes on another side of the Waard. The raging water flowed strongly to the southwest through a natural drainage, creating a channel now called Hollands Deep. Much of the Waard rapidly became shallow lakes or tidal waterways. At least twenty villages vanished and ten thousand people died, shattering community life. The once-prosperous city of Dordrecht, which thrived off the great polder, now stood on a tiny island. The disaster reduced many, including the local nobility, to penury. Anarchy descended on the surrounding countryside as raiders plundered granaries.

The Grote Waard had vanished beyond any hope of restoration. In its place was a wide area of open water. Generations passed before some limited reclamation work on its margins began.

Ultimately, the lack of an effective technology for land drainage and reclamation defeated the farmers, who had to drain their lands with gravity, scoops, or waterwheels. Salvation in the long term came with the invention of the wind-driven water mills and pumps, around 1408. Two centuries were to pass before such devices came into widespread use and ushered in the modern era of coastal sea defenses in the Netherlands.

Life along the North Sea shore was a constant tussle against the ocean, not so much against rising sea levels as attacks by unpredictable sea surges that coincided with exceptionally high tides. The farmers fought the attacking sea by armoring the coast with dikes and earthworks, the only viable defense in these landscapes. Human hands and backbreaking labor were often inadequate to the task, but humanly made sea defenses were the only long-term strategy for cities and towns at or near sea level. Armoring shorelines remains the strategy of choice against rising seas in many parts of the world. Twenty-first-century technology allows us to fence off the land at vast expense on a scale unimaginable even a century ago. The costs are gargantuan, beyond the purses of many endangered lands—and there is no guarantee that armoring will ultimately work.

"The Whole Shoreline Filled"

BY ABOUT 5000 B.C.E., Mediterranean sea levels had basically stabilized after the tumult of the Ice Age melt. Some adjustments to higher sea levels continued to persist, especially as rivers' flows became more sluggish and silt brought down by spring floods accumulated on coastal floodplains rather than being carried offshore into deep water. From the human perspective, the Mediterranean was, for all intents and purposes, a huge lake. The historian Fernand Braudel once called it a "sea of seas," where coastal trade in commodities and objects both rare and prosaic thrived. As early as the late tenth millennium B.C.E., farmers were crossing from Turkey to Cyprus. By 2000 B.C.E., hundreds of ships large and small plied eastern Mediterranean shores. The Minoans of Crete and the Mycenaeans of southern Greece were some of the most accomplished of these traders. And it was sailors from these ancient seafaring traditions that carried Greek warriors to Troy (also commonly known as Hissarlik) for the legendary siege immortalized by Homer in *The Iliad*, perhaps around 1200 B.C.E. But where did they land? Rising sea levels and the silting resulting from them, as well as earthquake activity, have altered coastal landscapes throughout the eastern Mediterranean beyond recognition.

Homer gives us the only description of the Achaean harbor during the siege of Troy, in about the twelfth century B.C.E.

[The Greek ships] were drawn up some way from where the fighting was going on, being on the shore itself inasmuch as they had

been beached first, while the wall had been built behind the hinder-
most. The stretch of the shore, wide though it was, did not afford
room for all the ships, and the host was cramped for space, therefore
they had placed the ships in rows one behind the other, and had
filled the whole opening of the bay between the two points that
formed it.[1]

Scholars have argued over the geography of the Trojan plain and the
location of the Greek camp for over two thousand years. The Greek
geographer Strabo wrote that "the Simoeis and Scamander [Rivers] effect a confluence in the plain, and since they bring down a great quan-
tity of silt they advance the coastline."[2] Strabo pointed out that the
coastline in his day was twice as far from the city as it had been twelve
hundred years earlier during Homeric times. Today the shallow bay below Hissarlik is dry land.

Both the Simoeis and Scamander would have had steep gradients
during the late Ice Age, when sea levels were much lower. As the ocean
climbed with global warming, so the river gradients became shallower,
just as they did along the Nile. Riverborne silt that had once flowed out

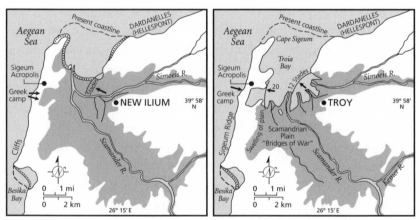

Figure 6.1 *Rising sea levels at Hissarlik (Troy): (a) the environs of*
Troy in Strabo's day and (b) the same area in Homeric times.

into the sea now arrived in the arms of much slower moving river floods. Instead of being pushed into deep water, the alluvium settled in shallow water. Currents in the Dardanelles also carried silt into the bay.

Core borings deep into the floodplain sediments have revealed details of the marine transgression that flooded the Trojan coastline after the Ice Age. Slowly accumulating clay and silt formed a delta floodplain. The cores show how the deepening silts slowly overrode the shallow marine muds that once filled what was, for a while, a bay and turned it into a plain. Constantly changing river channels dissected the new lowlands. Marshes expanded along the coastline, where sandy shoals close offshore protected the shoreline.

Comparing the geology with Homer's epic and Strabo shows remarkable agreement, with the Homeric Troia Bay extending somewhat inland of Hissarlik, with a distance of about four kilometers from the besieged city to the Achaean encampment on the Cape Sigeum Peninsula to the west, where the Aegean becomes the Dardanelles.

By Homer's time, maritime trade was big business in the Aegean Sea. Much of this commerce passed from village to village and bay to bay along the Turkish and Levantine coasts, with dozens of places of refuge

Figure 6.2 *Locations in chapter 6.*

within relatively easy reach, a few of them artificial ports built by human hands. The small port town of Phalasarna in extreme western Crete is a case in point. Three small islands protected the settlement, at a strategic location close to the eastern end of Crete and major trade routes. A rocky cape still provides shelter and good anchorage, with massive fortifications at the foot of the point. Immediately south of the fortifications once lay an artificial harbor accessible through two long canals, which are now dry. The port and canals have risen between six and nine meters as part of a general uplifting of western Crete, perhaps as a result of a major earthquake that struck the region about 365 C.E.

The earliest recorded settlement at Phalasarna belongs to the Minoans. According to a much later *periplus* (sailing directions) of about 350 B.C.E., the town, which by then was minting its own coinage, was "a day's sail from Lacedaemonia [the Peloponnese]." As late as the fourth century C.E., Phalasarna is said in another pilot book to have "a bay, a commercial harbor, and an old town."[3] At least four watchtowers guarded the harbor. Excavations have shown that there was enough water in the channels and the harbor for a warship with a displacement of 1.5 meters. Characteristic notches made by waves in rocks at the side of the channels tell us that the sea was at minimum 6.5 meters above today's sea level.[4]

In 67 B.C.E. a Roman praetor (field commander), Caecilius Metellus, destroyed pirate strongholds along the Cretan coast, for the islanders were notorious brigands equipped with fast ships. There are signs that the Phalasarna channel was blocked with large boulders, perhaps as part of this exercise, for the port with its watchtowers was too small be a viable mercantile port of call. Metellus may have sacked the town, but in any case, its days were numbered when a great earthquake raised the coast, perhaps within a few days, and Phalasarna became high and dry.

HISSARLIK WAS NOT ALONE. Silting has been a persistent consequence of rising sea levels throughout the Mediterranean. Again, we have Strabo as an authority, who described the Piraeus, the port of classical Athens, as "formerly an island . . . [that] lay over against the mainland from which it got the name it has."[5] Long before Strabo, centuries of oral traditions treasured by the Athenians referred to the Piraeus as an island. We know that a shallow lagoon connected the rocky island to the mainland during the fifth century B.C., when the Athenian general Themistocles, and also the statesmen Cimon and Pericles, built two long walls that connected the city with its two harbors, one of them the Piraeus. The walls turned Athens into a fortress connected to the sea, its source of prosperity and military power. But was the Piraeus an island when the wall builders got to work, or did they fill in a shallow lagoon that separated what was then an islet from the mainland?

Clearly the original island formed as a result of sea level rise, quite easy to chronicle in an area where earthquakes are rare. A team of geologists drilled ten boreholes in a search for answers.[6] They recovered not only layers of shallow water marine deposits, but also shell species that document a gradual change from island to mainland. So we now know that Piraeus was an island in the center of a shallow marine bay between 4800 and 3400 B.C.E., when the first civilizations came into being in Egypt and Mesopotamia. From then until about 1500 B.C.E., a lagoon separated from the sea by beach barriers lay between Piraeus and the mainland. Silt from the Cephissus and Korydallos Rivers gradually filled in the lagoon. Exactly when Piraeus became part of the mainland is still unknown, but it was some time after 1000 B.C.E. and certainly well before the long walls were constructed. Strabo, relying on oral traditions and himself an experienced observer of landscapes, was absolutely correct, just as he was at Troy. Here, as in the Dardanelles, river silt had slowly undone what rising sea levels had wrought and made it possible to turn Athens into a fortress with safe access to the sea.

Silting may have helped Athens, but it caused major problems elsewhere. Miletus was a port in western Turkey encircled by mountains. Minoans from Crete traded here as early as 1400 B.C.E. According to Homer, some Miletans fought against the Greeks at Troy. By the seventh century, Miletus had established over ninety overseas colonies, one as far away as the Egyptian delta, another in the Black Sea. Miletus later became a Roman city and continued to flourish until the fourth century C.E., when silt brought down by the Meander River finally closed the city's harbors and turned them into a swamp. Without its ports, Miletus was dead.

Slave labor and improved engineering saw the construction of the first large artificial harbors in places where there was little or no natural shelter. Tyre and Sidon, the great Phoenician maritime city-states, lay on the present-day Lebanese coast. The Tyre area has suffered from subsidence since ancient times, with the Phoenician port's northern breakwater lying about 2.5 meters below modern sea level. Here the water has risen at least 3.5 meters since classical times.[7] At Sidon, sea levels have been far more stable, with perhaps a rise of about half a meter. Thanks to core boring, we know something of the convoluted history of the two harbors. Between about 2000 and 1200 B.C.E., both cities relied on sheltered anchorages protected by natural promontories. Visiting ships lay in deep water, unloading their cargoes into smaller vessels to transport ashore. Sidon offered the somewhat better anchorage of the two, to the point that the city's population grew and there may have been attempts to build a harbor.

During the first millennium B.C.E., so many ships arrived at Tyre and Sidon that the Phoenicians built artificial harbors at great expense. Almost immediately, both the semiprotected ports suffered from silting, so much so that in later centuries Roman and Byzantine authorities had to dredge the harbors. After Byzantine times, the two ports were all but abandoned as Tyre and Sidon lost commercial importance. Both silting and coastal buildup submerged the harbors, which were lost until modern times, when archaeologists rediscovered them.

Another humanly constructed port also suffered from both silting and natural disasters. Herod the Great founded Caesarea on the present-day Israeli coast with a totally artificial harbor, said by the historian Josephus to be larger than the Piraeus, on a coast devoid of natural shelter in about 25 to 13 B.C.E.[8] It became the civilian and military capital of Judaea Province. Two moles of cemented pozzolana protected the harbor, each set in a concrete foundation. The pozzolana, a form of volcanic ash, came from Italy and required at least forty-four shiploads of four hundred tons apiece. The breakwaters provided adequate protection even from severe storms. Unfortunately, however, the often poorly mixed volcanic concrete did not bond well to rubble, which weakened them. Even worse, the port lay over a hidden geological fault line that runs along the coast. Earthquakes also ravaged the moles, causing them to tilt and settle into the seabed. Seabed studies have also shown that a tsunami attacked the area some time between the first and second centuries C.E. Whether this event merely damaged the harbor or destroyed it is unknown, but by the sixth century the harbor was silted up and unusable. Today, Herod's moles lie over five meters underwater.

THE ROMANS WERE industrious builders of harbors of all kinds. There were at least 240 major Roman ports in the eastern Mediterranean and around 1,870 in the west. Many such havens lay behind the coast on lagoons, in rivers, or even in canals.[9] Since most Roman ships were relatively small and rarely drew more than two meters, ports for even quite large cities could be small and shallow. Where there were no sheltered bays or natural anchorages as there were in the Aegean and Adriatic, the Romans turned to artificial harbors, built with slave labor. Their engineers had to confront the problem of silt, brought to the ocean by rivers large and small, then distributed counterclockwise by the Mediterranean current. For this reason, the Romans built some ports to minimize silting, among them Alexandria, which rose on the more protected, western side of the Nile delta.

The imperial Roman harbors at Ostia, which serviced Rome, lay to the north of the Tiber River.[10] Over the centuries, silt moving up the coast from the river clogged the artificial ports, which now lie several kilometers inland. The first major harbor, known as Portus, built by Emperor Claudius in 42 C.E., lay on the northern mouths of the Tiber. Two hundred ships sank here during a tsunami fourteen years later. The first Portus was not only vulnerable but also quick to silt up. Emperor Trajan finished a new harbor with a hexagonal shape capable of holding more than a hundred ships by 112 C.E. Ostia's harbors declined and silted up as the city became a country retreat for wealthy citizens of Rome. They finally fell into disuse during the ninth century after repeated attacks by Arab pirates.

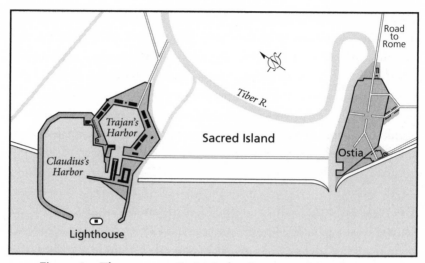

Figure 6.3 *The ports at Ostia, Italy. Courtesy of Southampton University.*

There were other harbors for Rome, too, all connected to the city by good roads. Each was a massive construction project. Pliny the Younger witnessed the building of Centumcellae harbor in the Bay of Naples by the Emperor Trajan, "where a natural bay is being converted with all speed into a harbor . . . At the entrance to the harbor an island is rising out of the water to act as a breakwater when the wind blows inland." Barge after barge brought in large boulders cast into the water until they formed a "sort of rampart . . . Later on piers will be built on the stone foundation, and as time goes on it will look like a natural island."[11] Trajan's port is now Civitavecchia, a dreary cruise ship port for Rome.

THEN THERE IS VENICE, the poster child of subsidence and rising sea levels in the Mediterranean. Venice began as a series of lagoon communities, a trading place founded by the city of Padua during the fifth century C.E. The settlements allied with one another in mutual defense against the Lombards, as the Byzantine Empire's influence dwindled. By the eighth century, Venice under its doges had become an important mercantile and shipbuilding center.[12]

Soon Venice's ships traded as far as the Ionian Sea and the Levantine coast. Venice became an independent city in 811. The city expanded closer to the Adriatic with canals, bridges, and fortifications, developing a close relationship with the ocean that was to endure for many centuries. Successive doges developed trade monopolies and acquired territories far from the Adriatic, with Venice reaching the height of its power in the fifteenth century. Despite constant wars and occasional reverses, Venice boasted of 180,000 inhabitants at the end of the fifteenth century, the second-largest city in Europe at the time, and certainly one of the richest in the world.

Over two million people lived under the Republic of Venice. Its wealth came from trade and shipbuilding, as well as fine textiles and jewelry, despite efforts by other powers to break its dominance in the eastern Mediterranean. The doges did everything they could to maintain strict neutrality between competing European monarchs. Venice remained wealthy, but coasted inexorably into slow decline, partly because of Portugal's rising dominance of the Asian trade. To add to its afflictions, vicious plague epidemics killed thousands of Venetians, a third of the city's population in 1630 alone. By the eighteenth century, the Adriatic was no longer a wholly Venetian world. Napoleon's soldiers occupied much of the Venetian state. The city became part of the Austrian empire in 1797, and part of Italy in 1866. Today it is capital of Italy's Veneto region, but only sixty thousand people live in the city itself.

Today the city extends over 117 small islands in the marshy Venetian lagoon along the Adriatic Sea, the lagoon extending along the coast between the mouth of the Po River in the south and the Piave River in the north. Venice's buildings stand on closely spaced, water-resistant alder woodpiles, many of them originally from Croatia. Most are still intact after centuries underwater, petrified by the constant flow of mineral-rich water that turns them into stone-like structures. They penetrate a layer of soft sand and mud before resting on hard clay. Even at low water, modern Venice is only just above the level of the Adriatic in high medieval times.

Venice oozes romance with its narrow canals and streets, gondolas, and magnificent palaces and bridges. But now it is also sinking, thanks to a combination of both rising sea levels (twenty-five centimeters over the past century alone) and subsidence caused by shifting tectonic plates and people drawing down groundwater. Not that the threat of floods is anything new, for flood tides push in from the Adriatic every year between fall and early spring, when storm surges generated by winter gales cause water to inundate much of the city. Six hundred years ago, the Venetians protected themselves from attacks from the mainland by diverting all the major rivers flowing into the lagoon. This draconian measure prevented silt from accumulating in the lagoon and filling in the shallows that protected the city, but made them more vulnerable to rising sea levels. Infrequent floods were just part of life during the winter months until the early twentieth century, when local industry began using artesian wells to extract water from subterranean aquifers for its water supplies. The city sank closer to sea level in response as subsidence intensified as a result at a time of rising sea levels. By 1996, ninety-nine floods inundated St. Mark's Square annually.

The great flood of November 4, 1966, caused by a record high tide behind a sirocco (southerly) wind, inundated the entire city. The waters lingered and caused much-publicized damage to priceless artworks. This was a wake-up call that led to long-term planning for protective measures. Well digging was banned during the 1960s, but the floods are still becoming more frequent, also often higher, inundating the ground floor levels of the city's buildings. Some studies claim the city is no longer sinking, but the economic stakes are enormous. Venice is now one of the major tourist destinations in the world, to the point that the magnificent city has become a glorified form of tourist theme park, with many fewer permanent inhabitants than in earlier times.

Since 1987, the authorities have embarked on an ambitious long-term project to safeguard the city and its lagoon. The MOSE Project (Modulo Sperimentale Elettromeccanico, or Experimental Electromechanical Module) is an ambitious attempt to check not only natural floods caused by strong winds that push waves into the Gulf of Venice, but also to slow erosion caused by the building of quays and other humanly built structures, as well as by the wakes of passing ships.[13]

Figure 6.4 *St. Mark's Square, Venice, during a high tide. Author's collection.*

The heart of this hideously expensive project is a system of seventy-eight mobile barriers designed to protect the three major entrances to the Venice Lagoon. The system consists of retracting, oscillating buoyancy flap gates that are designed not to interfere with the natural water exchange between the sea and the lagoon. The conditions surrounding the design are rigorous and include provisions that the barriers not impede fishing or navigation. The gates, which are large metal boxlike structures filled with seawater, normally lie on the bottom, supported by hundreds of concrete pillars. When a tide higher than 110 centimeters is forecast, the boxes are emptied using compressed air and rise, rotating around their hinges until they rise clear of the water. The gates then separate the lagoon from the tide within a space of about thirty minutes. The barriers comprise rows of gates, so that the operators have numerous combinations that can be used to accommodate different conditions. Locks, still currently under construction, will allow ships to pass through, one for larger vessels and two for small ones.

Seemingly monolithic and relatively inflexible, MOSE has been widely criticized by environmental groups, both on account of its expense and of the ecological effects it will have on the lagoon and marine life. For example, if the barriers remain up for some time, as seems likely in an era of more frequent inundations, then there is bound to be pollution in the now-nontidal water. The controversies surrounding MOSE have echoed through local and national politics for years, as well as involving the European Union, which considered the project to have serious polluting consequences for the lagoon.

Apart from these controversies, there remains a fundamental question. Will MOSE save Venice from ultimate submergence, given that St. Mark's Square is now underwater about a third of the time, as opposed to only seven times in the year 1900? The answer is surely no, for eventually flooding will reach an intolerable frequency and level a few decades from today, even if the barriers are in place. I can speak from personal experience. Carrying one's airline bags from the hotel to the ferry at the height of a Venice flood is a memorable experience, especially when hordes of aggressive tourists vie to dislodge you into the knee-deep inundation in St. Mark's Square. Such was my recent experience during a February high tide—not a record level, but one that lapped the foundations of the Doge's Palace and flooded the narrow streets. The high floods continue. In November 2012, 70 percent of Venice disappeared under floodwaters following a high tide and severe storm with heavy rainfall. High water in the city reached 149 centimeters, the sixth highest since 1872 and the fourth record flood since 2000. Tourists swam in St. Mark's Square. The authorities attributed the inundation to "global warming."

The water in the lagoon is rising about two millimeters a year, but recent studies had suggested that Venice had stabilized, that subsidence was no longer a problem. However, a new study by American and Italian researchers using GPS and space-borne radar measurements has documented a subsidence rate of about one to two millimeters annually between 2000 and 2010, probably caused by natural factors such as plate tectonics, which may be compacting the sediments under the city, rather than human activity. The islands in the lagoon are also subsiding at about the same rate or slightly faster. At the same time, the city is tilting very slightly to the east. If this research is confirmed, then the authorities will have to take into account subsidence when using the flood barriers. The researchers believe that the natural barriers that protect Venice may sink between 150 and 200 millimeters over the next four decades, unless the government reinforces the underlying sediments.[14]

What are the options if the floodgates are ineffective and subsidence continues? There is but one other solution—abandonment of the city. The huge sums of money required to move or jack up historic buildings and effectively rebuild the city in place on artificial foundations are simply not going to be available—given all the other calls that will arrive for assistance with rising sea levels from other Italian cities and elsewhere in the world. Many Venetians have already relocated to the mainland, but much more is at stake—the future of one of the world's historical treasures. The cost of moving even a few buildings is likely to be far beyond the capabilities of a sophisticated industrial society, even if the political will to save Venice is in place.

Venice is a sobering wake-up call, a prototype for what may happen to many low-lying cities with much larger populations in the not-too-distant future. Like some Pacific Islands or the Maldives in the Indian Ocean, Venice is a place in danger of vanishing from the face of the earth. There's a real chance that the historical theme park of today will be the underwater ruin of tomorrow. Even so, the plight of this much-treasured city pales alongside the vulnerabilities of twenty-first-century megacities, where millions will have nowhere to go.

"The Abyss of the Depths Was Uncovered"

Shortly after daybreak, and heralded by a thick succession of fiercely shaken thunderbolts, the solidity of the whole earth was made to shake and shudder, and the sea was driven away, its waves were rolled back, and it disappeared, so that the abyss of the depths was uncovered and many-shaped varieties of sea creatures were seen stuck in the slime . . . Many ships, then, were stranded as if on dry land, and people wandered at will about the paltry remains of the waters to collect fish . . . Then the roaring sea, as if insulted by its repulse, rises back in turn, and through the teeming shoals dashed itself violently on islands and extensive tracts of the mainland, and flattened innumerable buildings in towns or wherever they were found.[1]

The Roman historian Ammianus Marcellinus's dramatic account of the great earthquake that struck Alexandria on July 21, 365 C.E., understates the extent of the catastrophe. Thousands drowned; ships ended up on house roofs, some cast so far inland that they rotted where they lay. The disaster was so traumatic that Alexandrines commemorated the anniversary for two centuries.

Alexander the Great had founded what was then a small town on the western Nile delta coast in 331 B.C.E. Alexandria, named after the great conqueror, soon became one of the busiest ports of classical antiquity. The cosmopolitan city, with its more than three hundred thousand inhabitants, was a hub of commercial activity, a great center of learning famous throughout the Mediterranean world. Today, almost nothing remains of the harbor of two thousand years ago, which lies below sea level, the victim of natural subsidence.

Until recently, all we knew about the port came from descriptions by Strabo and other classical authors, and maps based on their writings over a century ago. Businessman turned archaeologist Franck Goddio has used electronic surveys of the now-sunken harbor to reconstruct the great harbor, with its breakwaters and famed lighthouse, the Pharos of Alexandria, one of the Seven Wonders of the Ancient World.[2] It is said to have been 140 meters high, a square tower with an iron fire basket and a statue of Zeus the Savior. An elaborate complex of palaces, temples, and smaller harbors lay within or close to the port. Subsidence in the order of five to seven meters since antiquity has effectively buried the port. Thousands of exploratory dives have yielded dozens of finds buried in the shallow water, many recovered with their exact positions marked by GPS, others, like the remains of a thirty-meter ship of the first century B.C.E., still left in place. Smaller artifacts include numerous magnificent artworks, well-preserved columns in pink granite from Aswan in distant Upper Egypt, and the remains of a palace said to be associated with Queen Cleopatra.

The Roman coastline has long vanished underwater. Six and a half kilometers offshore lie the remains of Heraklion, another port founded by Alexander, doomed by a combination of subsidence, earth movements, and shoreline collapse. Heraklion, the ancient Egyptian Thonis, lay on a peninsula between several port basins to the east and a lake extending to the west. The city with its bustling port controlled access to the now-vanished Canopic branch of the Nile. According to Herodotus, Heraklion was the mandatory port of entry for all ships arriving from "the Greek Sea."[3] Goddio and his colleagues have located an imposing temple to the sun god Amun, and also more than seven hundred anchors and twenty-seven wrecks dating to between the sixth and second centuries B.C.E. A large channel once passed through the city, linking the port to the lake and the river. The now-vanished port, finally destroyed by an earthquake during the eighth century C.E., was a gateway to Egypt for many centuries before Alexandria rose to prominence.

The fertile soils of the delta with its lagoons, marshes, and wetlands extended inland from Alexandria. They were a granary and the source of ancient Egypt's wines. Delta vintners nurtured fine wines, which enjoyed a gourmet reputation. The third-century Greek author Athenaeus, from the delta city of Naukratis, loved the wines from Mariut, southwest of Alexandria: "excellent, white, pleasant, fragrant, easily assimilated, thin, not likely to go to the head, and diuretic."[4]

RED WINE: the blood of Osiris, the god of resurrection. An early pharaoh, Scorpion I, went to eternity in 3150 B.C.E. at Abydos in Upper Egypt.[5] Three rooms of his sepulcher were veritable wine cellars, stacked with over seven hundred wine jars containing nearly forty-five hundred liters of fine vintages imported from the Levantine coast and from vineyards inland. Several centuries later, the pharaohs established their own royal winemaking industry to satisfy their desire for wines, often laced with figs. An Old Kingdom pyramid text tells us that deceased pharaohs dined off figs and wine, which it called the garden of the god. Such meals nourished the pharaoh in eternity.

Most royal vineyards were in the delta, begun with imported grapevines. The first vintners were foreigners. Even in later centuries, many were Canaanites, who brought sophisticated methods to the business from the beginning. The blistering heat of the delta required irrigation for water-sensitive grapes, but fortunately the soil was rich alluvium, nourished by the Nile flood and salt-free. Millions of amphorae of red and white wine from delta vineyards supplied pharaohs and temples over the centuries and provided drink offerings at major religious ceremonies. Tomb paintings show workers stomping the vintage, holding on to vines to maintain their balance. In 1323 B.C.E., mourning officials provided the young pharaoh Tutankhamun with amphorae of fine red and white wines labeled with the location and the names of the vintners, most of them from the western delta.[6] The reputation, and acreage, of delta wines survived into Islamic times.

Fertile soils nourished with silt and floodwater, flushed of any trace of rising salinity by the inundation: Ta-Mehu (the ancient Egypt name for the delta) was a patchwork of vineyards and productive cereal agriculture from very early times. The delta was Egypt's exposed northern flank, subject to inundation and to incursions from the sea, in human terms a permeable northern frontier, sometimes fortified and used as a military base for strong rulers, at others occupied by invaders from what is now Libya or from the Levant. Ta-Mehu linked the pharaohs with the ocean and the wider world beyond the horizon.

The pharaohs' greatest wealth was from agriculture, from the crops grown using the ideal natural cycle of the great river, much of the harvest coming from the delta. For more than three thousand years, the life-giving waters of the Nile defined ancient Egyptian life from the First Cataract to the Mediterranean. The river was an artery for transport and communication, the source of agricultural bounty, and it supported rich fisheries. Hardly surprisingly, the pharaohs developed a complex ideology that centered on the river and its annual inundation. Creation myths spoke of the waters of Nun, which receded to reveal a primordial mound, just like the floodplain emerging from the retreating inundation. The creator god Atum appeared at the same moment and sat on the emerging tumulus. It was he who was the first living being, who created order from chaos—just like every pharaoh who presided over the Two Lands.

In Ta-Mehu, as elsewhere in Egypt, three seasons defined the year: Akhet, the inundation, Peret, growing, and Shemu, drought.[7] The natural cycle of the Nile governed all attempts to cultivate the delta. Akhet provided life-giving water and silt, although human ingenuity could do much to improve the watering of the land. Farmers raised earthen banks to enclose large flood basins where they could retain water before releasing it. This was irrigation at its simplest, but it was usually sufficient to feed the four million to five million Egyptians who lived along the river at the height of the pharaohs' power during the second millennium B.C.E. There was no talk of cash crops in a self-sufficient agricultural economy. Such a notion was unthinkable to the pharaohs and their subjects. A small army of officials and scribes oversaw agriculture throughout the kingdom, but their interest was not in the nuances of farming, but purely in carefully documented yields for rents and taxes. For many centuries, ancient Egyptian villages fed the kingdom.

Farmers living in the delta may have labored in royal or temple estates as well as in village fields. There was generally enough to eat, crop yields were usually, but not invariably, more than adequate, and there was slow population growth with little or no threat from rising sea levels. Barrier dunes, extensive lagoons and marshes, and thick mangrove swamps served as natural fortifications for the farming landscape farther inland during thousands of years of very minor sea level change—most of that which did occur resulted from subsidence caused by earth movements.

WE SHOULD EMBARK on a minor environmental tangent here, for wetlands and marshes are important players in coastal protection in many parts of the world.[8] Wetlands are part of both aquatic and terrestrial environments. They are most widespread along low-lying coasts, such as those of the Florida Everglades, the Mississippi delta, or the delta shores of Bangladesh in South Asia. Like wetlands, freshwater and salt marshes are highly productive and often biologically diverse. Both provide invaluable protection against coastal erosion as well as providing excellent habitats for birds, fish, shrimp, and other organisms.

Salt marshes, flooded each high tide, have less biological diversity, for few plants can tolerate saltwater. They also trap pollutants flowing toward the shore from upstream and provide nutrients to surrounding waters. Being constantly inundated by rising tides, they are well adapted to rising sea levels. In the natural course of events, the marsh surface rises each year through the natural accumulation of sediments. At the same time, they also creep inland as sea level rises, maintaining themselves as the landscape changes, just as barrier islands do. Thus, they offer excellent natural coastal protection where the topography permits inland creep, provided agriculture and building do not destroy them, as now happens increasingly all over the world. Modern-day seawalls and other coastal defenses limit marsh expansion in estuaries, a chronic occurrence along the East Coast of the United States with its proliferation of coastal vacation properties. The destruction and especially restriction of inland expansion of marshes and wetlands have deprived us of vital coastal protection.

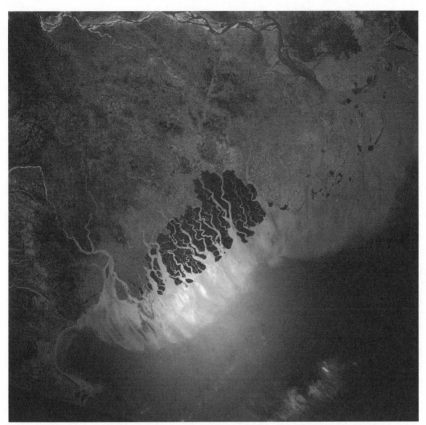

Figure 7.1 *Sundarban mangrove swamps (dark areas) lie at the mouth of the Ganges River in Bangladesh. Taken from the space shuttle* Columbia *on March 9, 1994. Courtesy: NASA.*

Mangroves protect coastlines over an enormous swath of the tropical world, including the Nile delta, in areas where waves are low and often where coral reefs protect them. Dense belts of them offer superb coastal protection, as the inhabitants of Homestead, Florida, discovered in 1992, when Hurricane Andrew's waves harmlessly surged into mangroves that protected the town. Mangroves also saved the Ranong area of Thailand from the great tsunami of 2004, which killed numerous victims in neighboring areas where the mangroves had been cleared (see chapter 10). Brazil is home to 15 percent of all mangrove habitats, but they line between

60 and 70 percent of tropical coastlines, or did until the twentieth century. Different mangrove species inhabit the diverse zones of coastal habitats, each growing at the limits of its tolerance for salinity. As the oceans warm, we can expect mangrove swamps to expand northward and southward from their present ranges, which is good news, for they nurture numerous species of birds, fish, reptiles, and shellfish. In Bangladesh, they were home to tigers. In Brazil and Colombia jaguars hunt prey at the edge of mangrove swamps.

For all their usefulness, mangrove thickets deter people with their tangled vegetation, numerous snakes, and endemic mosquitoes. They have inevitably become a target for destruction, cleared for agriculture and aquaculture, used as a source for house timber, a centuries-old industry in East Africa, whose swamps provided wood for houses in treeless Arabia. Today, the greatest threat to mangrove swamps comes from aquaculture, especially shrimp farming, which is a huge international industry in poverty-stricken countries like Bangladesh, Ecuador, and Honduras. Fortunately, mangroves are in no danger of extinction, for they are capable of adjusting effortlessly to sea level changes. Their greatest threat comes from humanity. We seem unaware that mangroves, marshes, and wetlands are some of our best weapons against onslaughts from the ocean.

BACK TO THE NILE: During the first millennium C.E., the delta rose to political and economic prominence, thanks to its burgeoning links with Greece, Rome, and other easily accessible, powerful kingdoms beyond Egypt's boundaries. The political equation changed decisively with the founding of Alexandria on the delta coast in 331 B.C.E., which brought the entire Mediterranean world to Egypt's doorstep. Much of the prosperity stayed downstream at a time when the diverse mouths of the Nile shifted and shifted again as a result of earthquake activity, subsidence, and coastal erosion. As was always the case until camel caravans became commonplace, most trade into the Nile valley traveled up the river. The delta beyond its growing cities and ports still remained somewhat of an agricultural backwater, despite its rich soils and excellent taxation potential, for there was negligible population pressure by modern standards. An estimated three hundred thousand people lived in Alexandria at the time of Christ, some half a million in Cairo a thousand years later, a medieval city that, thanks to Mamluk rulers, became a prosperous camel caravan meeting place after the thirteenth century and soon overshadowed the coastal city.

There was still plenty of cultivable and taxable land in the delta after Cairo rose to prominence. Repeated plague epidemics kept population growth in check, so it was still easy to adjust to sea level changes and coastal subsidence. Today the delta is so densely populated that there is not enough land to go around and the ocean is moving inland. Compare the medieval figures of 300,000 and 500,000 people with those for 2011: Over 3.8 million people crowd into modern Alexandria, while Cairo with its over 6 million urban inhabitants has a staggering population density of 17,190 people per square kilometer, one of the most densely populated cities in the world, this before one factors in the surrounding suburbs. The soaring urban and rural populations of both the delta and Upper Egypt have brought the complex relationship between the Nile and the ocean to the brink of crisis.

The present delta crisis has its origins in the early nineteenth century, when the Mamluk Muhammad Ali Pasha, well aware of the economic potential of Lower Egypt's soils, decided to increase cotton production in the delta, cotton being a lucrative cash crop. Ali deployed forced labor from villages along the Nile in an orgy of deep canal digging.[9] His ruthless practices and the ambitious, often unsuccessful projects of his successors resulted in massive international debts and a breakdown of political order that led to British occupation in 1882. The new occupiers in turn invested heavily in irrigation agriculture both as a way of paying off foreign creditors and to feed a burgeoning population, over seven million in 1882 and up to eleven million by 1907. Even with these investments, there was insufficient inundation water to go around. The British consul general, Lord Cromer, laid plans for a network of dams and barrages to expand irrigation and to reduce dependency on unpredictable floodwaters.

His timing was impeccable politically. By this time, epic Victorian explorations had traced the Nile to its multiple sources. Furthermore, the British controlled the entire length of the river except the upper reaches in Ethiopia, so long-term planning and major dams were practicable for the first time. The so-called Aswan Low Dam was constructed immediately below the First Cataract between 1898 and 1907, with numerous gates to allow both water and silt to pass through.[10] The dam was too low, so it was raised twice, in 1907–1912 and again in 1929–1933, giving it a crest thirty-six meters above the original riverbed. Even that was not enough. After the inundation nearly topped the dam in 1946, the British decided to build a second dam about seven kilometers upstream. Changing political circumstances and international maneuvering delayed construction until 1960–1976. The massive rock-and-clay Aswan Dam, 111 meters tall and nearly 4 kilometers from bank to bank, truly blocked the river, forming Lake Nasser, a huge reservoir 550 kilometers long.[11] Spillways allow a constant flow of water through the barrier, so the Nile is no longer a natural watercourse but effectively a huge irrigation channel. Instead of overflowing its banks, the river remains at the same level year-round.

The Aswan Dam protects Egypt from floods and the immediate effects of drought, but the long-term environmental consequences are only now beginning to come into focus. In predam days, when the Nile flooded, the low groundwater level when evaporation was highest in summer prevented salt rising to the surface through capillary action. Now the groundwater level is higher and more constant natural flushing has virtually ceased. As a result, soil salinity has increased dramatically, endangering agricultural yields. Only now are large-scale, and extremely expensive, drainage schemes attempting to take care of the problem. Before the Aswan Dam, 88 percent of the silt carried downstream ended up in the delta. Now 98 percent of the silt load remains above the dam and virtually no silt reaches the ocean. Chemical fertilizers have replaced silt; diesel pumps rather than gravity provide irrigation water year-round. Some years the authorities close off the dam for two or three weeks in winter. Far downstream, the surviving delta lake levels fall rapidly; seawater promptly enters them.

Figure 7.2 *Constructing the first Aswan Dam around 1899. D. S. George/British Library.*

Even the construction of the Aswan Low Dam affected the distant coastline, for reduced silt levels removed some of the natural barriers to coastal erosion. The High Dam has had much more widespread effects that may even be felt far from the delta, elsewhere in the eastern Mediterranean.[12] Groin and breakwater construction to refresh beach sand along popular beaches has had some beneficial results, but the long-term effects are impossible to discern. There are other problems as well. With many fewer nutrients carried to the coast, both lagoon and in-shore marine fisheries are fading rapidly. Sardines off the delta depended heavily on phytoplankton during the flood season. Absent the inundation, the Egyptian sardine fishery declined from about eighteen thousand tons in 1960 to around six hundred tons in 1969, even less today. Stocks have now recovered considerably, largely because of winter outflows from coastal lakes, but there are signs that sardine migration patterns have changed. Water hyacinths are clogging waterways and canals, increasing water loss through evaporation and transpiration (loss of water vapor from plants).

Quite apart from the ecological damage, escalating demands for water for industrial, urban, and agricultural purposes upstream of the delta mean that less and less of the Nile reaches the northern reaches behind the shoreline. Much of what does is polluted by industrial waste, municipal wastewater, and runoff from fields treated with heavy doses of chemical fertilizer. The polluted water flows into already shrinking coastal lagoons, threatening fisheries and waterfowl habitats.

All this is before one factors in population growth and inexorable sea level rises associated with global warming. Egypt's population is expected to exceed a hundred million people by 2025. Sea levels will rise by at least a millimeter a year into the foreseeable future without accounting for projected climate changes. Combine these sea level figures with estimates of the effects of natural land subsidence, commonplace, as we have seen, in the eastern Mediterranean, and the forecasts are even more depressing. Sophisticated computer models project a relative rise of sea level of between about 12.5 centimeters to an extreme of 30 centimeters in the northeastern delta. These are by no means vast rises in vertical terms, but they are potentially catastrophic horizontally across the northern delta, where the relief varies by little more than a meter. An estimated two hundred square kilometers of agricultural land will vanish under the ocean by 2025, at a time when the rural population density of the delta is rising rapidly.

At present, the Mediterranean is winning the long standoff between land and sea. More sediment now leaves the delta than arrives there. As a result, the ocean will cut back the shore vigorously, even at points of resistance like promontories. Sand washed ashore by storms will infill coastal lagoons. Constantly shifting dunes will form in many places, which will be hard to stabilize by any form of artificial protection. Coastal sand barriers will migrate inland, filling the remaining coastal lagoons, which are already subdivided by expanded irrigation work, roads, and other modern industrial infrastructure. As marshes and swamps disappear, already decimated reserves for migrating birds and

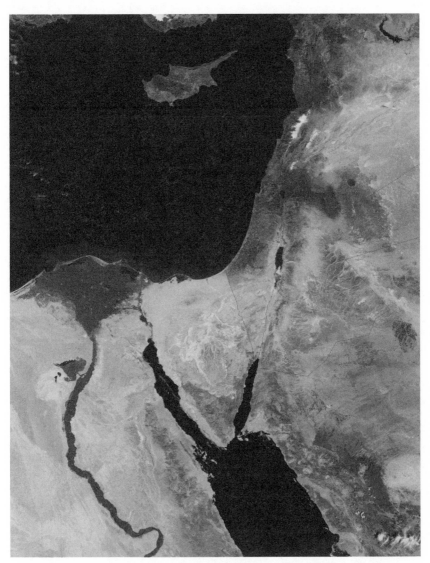

Figure 7.3 *The Nile River valley and delta from space, an image taken on September 13, 2008, with MODIS, NASA's Moderate Resolution Imaging Spectroradiometer aboard the Aqua satellite. Courtesy: NASA.*

waterfowl will vanish. By 2025, inexorable pumping will encourage the inland migration of saltwater-laden groundwater at a time when more and more land is being diverted from agriculture to urban and industrial use.

The delta has ceased to function as a balanced system. What can be done to solve these daunting problems of land shortage, groundwater pollution, and environmental degradation resulting from an exploding population and increasingly scarce water supplies? In many respects, the Nile delta is akin to the Netherlands, a land trapped between a rising ocean and the land. Egypt does not have the financial resources to embark on a massive coast protection project like those along the North Sea. Nor are there the funds to construct the artificial wetlands and treatment facilities for recycling wastewater—or regenerating mangrove swamps to protect the coast. To restrict and control the increasingly limited waters of the Nile will require strong political will, and also an infrastructure and mechanisms for ensuring fair shares for all, from individuals to industry and agriculture. The stakes are literally life and death. One estimate has it that food shortages resulting from environmental degradation and rising sea levels will trigger massive famines that could turn more than seven million people in Egypt into climate refugees by the end of the twenty-first century–and that figure is conservative. Herodotus was right when he wrote that the Egyptians would suffer if the Nile flood dried up, for drying up it is just as sea levels rise inexorably.

"The Whole Is Now One Festering Mess"

Lothal, Gujarat, India, 2100 b.c.e. The rectangular dockyard bakes under the summer sun. Densely packed cargo ships with battered, sewn planked hulls lie cheek by jowl alongside the quays. Sweating laborers in white loincloths heft bundles of cotton over narrow gangplanks and toss them into empty holds. The tide is high, so the lock gates are open. A heavily laden vessel moves slowly into the narrow defile, her crew straining against long poles. She clears the lock; the stone gate closes. The steersman catches the first of the ebb with his oar and pilots his clumsy charge toward the open sea. High above the port, white figures gaze down from the mud-brick citadel that broods over the bustling harbor.

Lothal, eighty kilometers southwest of the modern city of Ahmadabad, was an important, almost unique port on the western Indian coast three thousand years ago.[1] Its anonymous rulers, their names lost to history, commanded the navigable estuaries of the Sabarmati and Bhogavo Rivers where they emptied into the narrow Gulf of Khambhat (Cambay). At the time, the ocean was a mere five kilometers from the prosperous city. From Lothal, cargo ships passed far upstream up the two rivers. Stone anchors in the now-dried-up river channels have been unearthed at least fifty kilometers inland. Today, massive silt deposits laid down by river floods and ancient storm surges have isolated Lothal from the ocean with its predictable monsoon winds that once gave it sustenance.

No one knows when farmers first settled at Lothal, but it was at least five thousand years ago, when a small agricultural settlement rose on a low mound on higher ground on a wide alluvial plain protected from river floods by a mud dike. The village gradually became a town and craft center, well known for its stone-bead workshops, and also for lustrous red pottery, made from a glittering micaceous clay. By 2400 B.C.E., the growing town was part of a much larger interconnected world, that of the Harappan civilization, a tapestry of cities, towns, and villages centered on the Indus and Saraswati Rivers to the north.[2] Lothal's appeal came from the fertile soils in its hinterland, which were ideal for irrigation agriculture and cotton cultivation, the latter an important export to distant lands. The city itself became not only a manufacturing center but also one of the major ports that linked the cities of the Indus Valley to other lands.

The monsoon winds of the Indian Ocean are among the most useful of all prevailing breezes, for they obligingly blow in opposite directions seasonally.[3] The winter brings the gentle northeast monsoon, the summer the more boisterous, less predictable southwesterlies when the west coast of India forms a lee shore and is best avoided. As a result, a ceaseless rhythm of sporadic voyaging developed over the centuries. These were coasting passages that passed from village to village, bay to bay, between the Gulf of Oman and the Persian Gulf and lands perhaps as far away as Southern India's Malabar Coast, where shipbuilding timber could be found. Cotton, stone beads, fine pottery, and other goods passed east. Gold, silver, copper, and other commodities came from ports in the west, perhaps in modern-day Bahrein. This coasting trade never assumed the dimensions of later long-distance Indian Ocean commerce, for the Indus Valley cities were oriented more toward the land and terrestrial trade networks than to the ocean. Nevertheless, Lothal became the most important port in their world. The mouth of the Indus River to the north ended in an extensive delta of braided channels near Karachi that presented formidable navigational challenges with its shallows. According to an Egyptian-Greek pilot of the first century C.E., "The water is shallow, with shifting sandbanks occurring continually and a great way from the shore." Numerous sea serpents and "a very great ebb and slow of tides" added to the hazards.[4]

At first ships visiting Lothal had to tie up alongside a primitive quay in fast-moving tidal waters. When the rivers overflowed their banks in 2350 B.C.E. and destroyed the town, the port's leaders transformed the strategic ruined location into an imposing metropolis where the principal buildings stood above flood level on brick platforms. The entire undertaking took years and involved building an artificial dockyard thirty-seven meters long, twenty-two meters wide, and four and a quarter meters deep. An inlet at the northern end with gates connected the dockyard to the Sabarmati River at high tide—the earliest known lock in history. A 260-meter-long wharf on the east side of the dockyard connected it to a huge warehouse built on a platform over three meters above flood level. Beyond the docks lay the upper town, also built on a platform, where the rulers and prominent individuals built imposing houses. A short distance away was the lower town with its workshops, bead makers, and potters. The entire city boasted of an elaborate drainage and sewage system.

For all its thriving export trade, Lothal was never a completely safe port. Two unpredictable forces were in play: river silt and the occasional

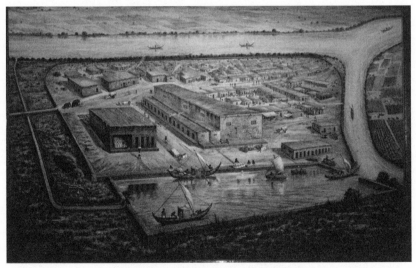

Figure 8.1 *Artist's impression of the Lothal dockyard. Archaeological Survey of India/Omar Khan Images of Asia.*

tropical cyclone. The city lay on a slightly elevated portion of river floodplain, formed by silt carried downstream by monsoon floods between July and September. Fine silt made for fertile soils; irrigation agriculture worked well for cotton and other groups, except in years when the monsoon rains were exceptionally heavy. Such rainfall tended to arrive in La Niña years in the distant southwestern Pacific Ocean, when the southwestern monsoon was more powerful than usual. Such rains bred catastrophe. Swollen rivers burst their banks. Silt-laden floodwaters submerged villages and destroyed irrigation canals and growing crops. On many occasions, thousands perished both from the inundation and from the famine and epidemics that ensued. The same hazard menaces much of Pakistan and the Lothal region today. In 2010, monsoon floods in the western Punjab and Sindh affected over 570,000 hectares of cropland and inundated a million dwellings. Over two thousand people perished. Another major flood in 2011 left two hundred thousand people near Karachi alone homeless. Such monsoon floods in the past would have had devastating effects far downstream at Lothal, where rapid silt buildup and rushing floodwaters could inundate much of the city.

The flood threat increased gradually as the river mouth silted up over three centuries. All the authorities could do was to raise house foundations and change the levels of municipal drains. One particularly destructive flood even damaged the thirteen-meter-thick platforms that supported the upper city. Many buildings in the lower town collapsed; streets filled with flood debris; floodwaters breached the embankment walls of the dockyard. Community leaders quickly deployed large numbers of people to clear drains, rebuild houses, and expand the city. Lothal reached the zenith of its prosperity in 2000 B.C.E., so much so that the people apparently became complacent and allowed flood defense works to fall into disrepair. At this point, the second villain appeared on stage—a massive tropical cyclone.

Severe tropical cyclones tend to form in the north Indian Ocean, on either side of the Indian Peninsula, between April and December, with peaks in May and November.[5] Major cyclones are relatively uncommon in the Arabian Sea, but those that do occur bring very strong winds, produce high waves, and, above all, spawn powerful sea surges. In June 2007, Cyclone Gonu, a Category 5 storm with 258-kilometer-an-hour winds, descended on Oman at the mouth of the Persian Gulf, only the third such storm to hit land there since 865 C.E. Fortunately the narrow mouth of the Gulf weakened the storm, which caused widespread flooding and loss of life in both Oman and the United Arab Emirates. At the time when Lothal was a major port, sea levels in the Gulf were higher, which would have made it possible for a cyclone to surge as far north as southern Mesopotamia.

In the absence of modern forecasting tools, predicting cyclones far ahead of their arrival is virtually impossible, beyond telltale cloud effects, rising humidity, and strengthening gusty winds some days and hours ahead. The damage wrought by winds blowing at 120 kilometers an hour or more is severe enough, but it pales into insignificance alongside the sea surges, which sweep everything before them and can raise sea levels by twelve meters without warning. The surge may last only a few hours, but the damage is done. The destruction is particularly severe when the surge coincides with a high tide.

Cyclones have a long history in the Arabian Sea, but unfortunately we have no record of them before scientific records began in the nineteenth century. Many of the floods that afflicted Lothal and other coastal cities resulted from exceptional monsoon rains, but the most severe blows came with cyclonic storm surges such as one that swept ashore in about 2000 B.C.E. Such events are entirely different from monsoon floods, when floodwaters rise more slowly and there is at least some warning of pending trouble. Four thousand years ago, the powerful surge arrived suddenly. Raging water submerged the Acropolis and warehouses, and also entire quarters of the city. Lothal effectively ceased to exist within a few hours. Overseas trade came to a standstill; many of the inhabitants fled permanently to higher ground.

Only a small settlement rose on the submerged ruins of Lothal. The survivors repaired the dockyard embankments, but they now had to dig a two-kilometer canal to link the port to the river. Even then, only small ships could be locked through the inlet. Larger cargo vessels had to moor alongside a quay in the river. The volume of overseas trade was now so small that the city declined into obscurity and lost contact with the great Indus cities. Ironically, the floodwaters accumulated a five-meter-high mound that seemingly rendered the ravaged city immune from further flood damage. The natural defense was an illusion. Another massive sea surge a century later obstructed the normal flow of the spring tides, choked the natural drainage system, and turned the entire estuary into a huge lake. Villages, towns, irrigation works, and dams vanished before the deluge. This time the dockyard disappeared under a huge silt deposit, never to be restored. All political authority appears to have evaporated. A rectangular city protected from flooding by a thirteen-meter brick wall finally vanished forever. At the time, the ocean was only few kilometers from Lothal. Today, silt buildup resulting from flood and rising sea levels means that it is now over twenty kilometers inland.

We have only a dim understanding of the complex geological forces that contributed to sea level changes along the western Indian coast. We know that coastal uplift along the northern margins of the Arabian Sea as well as earth movements may have contributed to the vulnerability of Lothal and other coastal settlements. The same general conditions affected later societies living along the western coast, but the main destruction stemmed not from sea level rise or subsidence, but from heavy monsoon rains and from extreme weather events—tropical cyclones. As they had in earlier Harappan times, farmer and city dweller alike lived in a monsoon environment, where the intensity of the seasonal rains made all the difference between famine and plenty. The failure of the monsoon, often associated with strong El Niños in the southwestern Pacific, could kill hundreds of thousands of farmers, even millions, as happened in the late nineteenth century with the epochal western Indian famine of 1878.[6] A strong La Niña condition, the opposite swing of the Pacific climatic seesaw, could bring exceptionally heavy rains. Such deluges were just as destructive as drought, especially in coastal regions where ponding and silt accumulation could destroy valuable irrigation works and farmland.

The destruction wrought by extreme weather events along the western coasts of what are now India and Pakistan over centuries and millennia can only be guessed at. All we have as a basis for comparison is historical

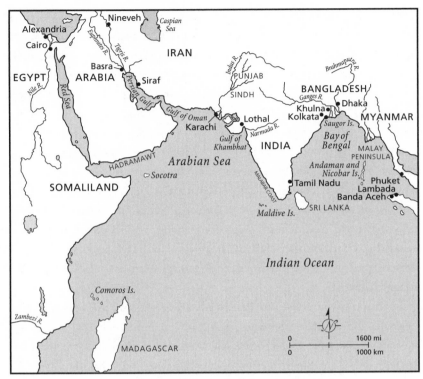

Figure 8.2 *Map showing locations in chapter 8.*

and modern experience in Bangladesh at the head of the Bay of Bengal on the other side of India, where millions of people live only a few meters above sea level and farm at the mercy of monsoon rains and the capricious violence of tropical cyclones.

ABOUT 80 PERCENT of Bangladesh is alluvial floodplain, most of it less than ten meters above sea level, the world's largest river delta formed by the Ganges, Brahmaputra, and Meghna Rivers. Water covers some ten thousand square kilometers of the country. Much larger areas flood during the monsoon season. Only some landscapes far inland have more rugged topography, bisected by fast-flowing rivers. The delta coastline extends over about six hundred kilometers facing the Bay of Bengal.

Tidal action molds the shoreline, where an arabesque of waterways large and small cuts through the coastal plain, the largest of them about half a kilometer wide. Strong tidal streams keep the river channels long and straight as they flow through clay and silt deposits that resist erosion. Easily eroded sand forms banks and small islands at the river mouths, which strong southwesterly monsoon winds convert into dunes above the high-water mark. Mudflats of fine sediment form behind them and eventually become islands. This dynamic environment, a unique mosaic of beach and tidal forests, and also dense mangrove swamps, acted as an important cushion against sea surges and cyclones. Two hundred years ago, more than eleven thousand square kilometers of mangrove swamps and forests protected the coast. Today this natural coastal barrier is under threat from promiscuous forest clearance for agricultural land, for shrimp farming, and to construct barrages for irrigation works.

Right up to the coast, the delta soils are fertile and ideal for all kinds of crops, among them rice and cotton, so it is hardly surprising that numerous farmers live close to the sea. Unfortunately, their homeland lies at the head of the Bay of Bengal, which narrows like a funnel toward its northern coast. Fierce tropical storms with their accompanying sea surges form in the open sea, then barrel northward, gathering strength as they approach the coast. A shallow continental shelf heightens the surges as they move inshore, making them higher than in many other parts of the world.

Cyclones and their fearsome sea surges descend on Bangladesh with a furious intensity that has killed millions of people over the centuries. Between 1947 and 1988 alone, thirteen severe cyclones have ravaged the lowlands causing thousands of deaths and sweeping away villages and defensive embankments. Over the next century, projected sea level rises of a meter or more will intensify the maritime siege.

Countless sea surges have attacked the low-lying Bengal coast since Lothal in the west was in its heyday four thousand years ago.[7] This may be why flood stories loom large in ancient Indian religious works like the Hindu Satapatha Brahmana with its myths of creation and of a huge flood, compiled between 800 and 500 B.C.E.—an equivalent to the

Great Flood in Genesis.[8] Monsoon floods were a routine part of the seasonal round, even if they were unusually large, whereas a huge storm surge driven by a cyclone was something to be remembered beyond the span of a single generation. Unfortunately we have no accurate records of cyclonic events until the nineteenth century, but there can be little doubt that the sequence of meteorological events was much the same far into the past. We know, for example, from Indian records that in 1582 C.E., a five-hour cyclone and associated sea surge killed an estimated half million people in what was then Bengal. Only Hindu temples with strong foundations survived the attack.

There are listings of eighteenth- and nineteenth-century storms, but there is little information on casualties or damage until the founding of the Indian Meteorology Department in 1875, which came about as a direct result of the severe cyclone of 1864, the first to be recorded in any detail. The report on the storm came from a plethora of meteorological observations, official reports, letters, ships' logs, and firsthand accounts. By modern standards, it's a patchwork of impressions, but it is brought to life by eyewitness reports, written or recorded when memories were horrifyingly fresh.[9]

The weather was blessedly dry in Kolkata (Calcutta) on October 4, 1864. After five months of heavy monsoon rains, there had been no showers for several days. The air felt clean and dry with less humidity than usual. British officials and local people alike reveled in the calm, pleasant day with its light winds. No one knew that the barometer was falling sharply to the south in the center of the Bay of Bengal. A cyclone had formed over a circular area about 650 kilometers across. Unsettled, squally conditions affected a wide area, including the Andaman Islands. Ships' logs hint at rough weather as early as October 3. The steamship *Conflict* experienced "frightful squalls with very heavy rain." A day later, the storm center advanced toward the mouth of the Hooghly River and in the direction of an unsuspecting Kolkata 130 kilometers upstream. "Furious squalls from the north, with torrents of rain" hit the northward-bound steamer *Clarence*. Strong northerly winds, high seas, and heavy rain showers lashed the Bengal coast. The pilot vessel at the mouth of the Hooghly slipped her anchor cable and ran offshore for

Figure 8.3 *Ships on the Hooghly River, Kolkata (Calcutta), after the 1867 cyclone. Samuel Bourne/British Library.*

safety. Vicious waves washed away her quarter boat and threw the ship on her beam-ends until the mainmast was cut away.

The next day the storm center passed over the coast. A tug anchored off the Saugor lighthouse at the Hooghly River mouth steamed at full power into the teeth of the wind while still at anchor. The cable parted; the wind dropped for three quarters of an hour as the storm center passed overhead, then resumed with savage gusts that blew the vessel onto its beam-ends, "burying her in a sheet of foam to the top of the funnel." As the wind finally dropped, the tug went aground off an unfamiliar shore and dried out at low tide. Fortunately the captain was able to back off as the water rose and anchored safely.[10]

This cyclone was probably a Category 4 storm on today's Saffir-Simpson Hurricane Scale, weakening to Category 3 as it advanced over the land. The violent winds caused a great deal of damage near the Hooghly's mouth, blowing down trees and flattening houses. The center passed about forty kilometers west of Kolkata. Overall, the damage was severe but not catastrophic until a huge storm surge moved up the Hooghly at about ten A.M. on October 5, two hours before high water. The surge advanced slowly up the river, arriving in Kolkata only about an hour before high tide, so the overall effect was extreme. The water rise may have been as much as twelve meters. A huge accumulation of water arrived in Saugor Bay at the river mouth where the steamer *Martaban* lay at anchor waiting to enter the estuary. The raging wind and surge dismasted her and wrecked everything on deck. She drifted helplessly across some of the most dangerous shallows in the mouth of the river. Astonishing though it may seem, the crew were able to take soundings through the height of the storm. There was never less than about thirteen meters below the keel where normally there was virtually no water at all. The survey ship *Salween* was also anchored off Saugor Island. When the cyclone struck, she soon dragged her anchor in the hurricane force winds. The captain slipped the cable and sailed into deeper water under one small staysail. Early on the morning of October 5, the ship grounded on the "beach opposite the Post Office." Her sails were in tatters, the lifeboat washed away, great waves breaking over the deck. The ship's log tells the tale: "Observed that the storm-wave was carrying us in shore, as we passed over the tops of several trees." When the water receded, "found the ruins of the Telegraph Office under our jibboom."[11]

The storm surge arrived without warning in an overpowering advance. An official traveling in the lowlands took shelter in a convenient hut. He described "a most curious sound which was exactly like the letting off of steam from a steamer, but on a gigantic scale." The cyclone ripped the roof off the hut and collapsed its walls. As the official peered anxiously through the driving rain, the "water all at once suddenly rose as if by magic, and steadily rolled towards us . . . The water reached us exactly at 10 minutes to 1, at which hour, being up to my waist, my watch stopped."[12] The survivors clung to two coconut trees until the wind dropped.

Throughout western Bengal, terrified villagers took to their huts, huddling in the darkness with only the roaring wind as company. As the surge rolled up the Hooghly without warning, it flooded all low-lying terrain for up to sixteen kilometers on either side, sweeping everything before it in a sudden rush of water. Entire villages vanished as the water broke through dikes and breakwaters, leaving no signs of human habitation in its train. Houses vanished, their owners drowned in a moment. A few survivors managed to cling to floating roofs or pieces of wood, only to be carried helplessly for kilometers. Cattle, even tigers, were swept along, prey and predator alike. Only water buffalo, always strong swimmers, were able to master the rapid currents. The local district superintendent of police, one R. W. King, interviewed the few survivors, reporting that women and children fared worst, trapped as they were in their dwellings thrown down by the shrieking wind. He added, "The whole is now (2nd November) one festering mess. I attempted to go near it, but the fearful stench rendered it impossible for any one to do so. The only course to follow with this and other villages similarly circumstanced, will be to leave them until the dry weather, and then to fire the whole mess."[13]

The morning after the storm, the Hooghly was awash in corpses and dead beasts. A foul miasma from unburied bodies permeated the air, many of them of young children. Those who survived had no food; rotting vegetation and seawater had polluted their water tanks. In desperation, they ate spoiled food and drank impure water. Cholera, dysentery, and smallpox raged as villagers starved. No one knows how many victims died in the cyclone and its surge, but at least fifty thousand lost their lives in the storm wave, while an additional thirty thousand succumbed to disease. Entire regions lost more than three quarters of their population; at least one hundred thousand cattle drowned; saltwater turned the rice crop black in the fields.

There were 195 ships in port at Kolkata, either alongside the docks or anchored in the river. This far upstream, the surge was considerably reduced, thanks to the friction of the river bottom and the loss of water over the riverbanks. The authorities had laid exceptionally strong moorings as a result of another cyclone in 1842, but the mooring chains were too short for the surge and ripped the anchors from the bottom or pulled ships underwater. Only twenty-three ships escaped damage, a small consolation for London shipping companies saddled with heavy losses. Most ended up in a confused mass of cargo ships, lighters, and small boats cast up on nearby sandbanks. Bundles of jute packed for export littered the riverbanks. Looters were rampant and many became rich overnight. One two-thousand-ton Peninsular and Oriental liner, the *Bengal*, remained aground for two months. Digging a dock around her at vast expense refloated her. A conservative estimate of the total damage to the port and shipping lay at about a million pounds sterling (about seventy-seven million dollars today).

Another epochal storm, the Great Backerganj Cyclone of October 29, 1876, brought winds of 220 kilometers an hour and a ten-to-fourteen-meter sea surge. Unsettled weather followed several days so hot and still that pitch boiled out of deck seams of ships bound from Sri Lanka to Kolkata.[14] Thick gray clouds and heavy rain accompanied the approach of the storm, which caught many vessels making their way up and down the Hooghly River. The steamship *Penang* was steaming down the river into the teeth of the now-vicious cyclone. Huge waves broke aboard and flooded the saloon. The engineer and his crew were battened down below, the rig and deckhouses completely shredded. The ship "lay like a log," her saloon flooded.[15] When the wind subsided, she staggered back to Kolkata, a virtual wreck on deck. Other ships suffered massive damage and dismastings. Some went aground; others were cast on their beam-ends. The most catastrophic damage came from the sea surge, which again followed on a high tide. The pressure exerted by the storm prevented the tide from ebbing, so the storm waters stalled over the shallows near the mouth of the Meghna River. Once the surge overpowered the tide at about two A.M. on October 31, the water covered low-lying coastal areas and islands to a depth of as much as twelve meters in less than half an hour. By eight A.M., the water had receded, but the destruction was universal and even more catastrophic than that of 1864. At least a hundred thousand people drowned and a further hundred thousand perished from cholera and related diseases. The total casualties approached a quarter of a million.

AT LEAST SIX recorded cyclones have killed over a hundred thousand people. One of the worst storms on record came on November 13, 1970, when Cyclone Bhola hit the entire Bangladesh coast.[16] The storm formed over the central Bay of Bengal, then traveled north and intensified, with

winds as high as 185 kilometers an hour. The cyclone's storm surge devastated islands off the Ganges delta, leveling entire villages and destroying crops over a wide area. At least half a million people died, and
also a million head of cattle. More than four hundred thousand houses
vanished and thirty-five hundred schools and other educational institu-
tions were inundated. Local fisheries suffered heavy losses. Forty-six
thousand fishermen perished, with about 65 percent of the fishing capacity of the coastal region being destroyed, in an area where 80 percent of
protein comes from fish. Three months after the storm, three quarters
of the local population was receiving food aid. The storm affected over
three and a half million people with various degrees of severity. Studies
after the cyclone revealed that about half the casualties were among children under ten years old. The Bhola disaster prompted Beatle George
Harrison and Ravi Shankar to organize the Concert for Bangladesh, the
first such relief concert for aid, in 1971. The relief concert coincided
with the formation of an independent Bangladesh and a new era in
cyclone forecasting. Unfortunately, as we shall see in chapter 11, both a
rapidly growing population and accelerating sea rise have complicated
the equation.

Backerganj and Bhola—two great cyclones of the past two centuries
offer a daunting portrait of human vulnerability to sea surges and other
catastrophic events that affect low-lying coastlines. They also reinforce
an often inconspicuous reality of today's world. Millions upon millions
of us dwell, often in densely populated megacities, at the edge of the attacking sea. We are vulnerable to the ocean and its whims in ways unimaginable even one or two centuries ago.

The Golden Waterway

HEMUDU, YANGTZE RIVER DELTA, CHINA, CA. 4500 B.C.E. A savage typhoon wind booms overhead like a prolonged, rolling thunderclap, swishing through the reeds with frightening power. Gray clouds roil over the shallows; blowing dust envelops the planked houses close to the rice fields. Normally typhoons give warning of their approach—mounting gray clouds, still winds, and intense humidity. Every farmer knows the menacing signs. But this fast-moving storm has arrived unexpectedly, so there is no time to move to higher ground beforehand. The villagers shelter inside their dwellings, their portable possessions bundled close at hand. Suddenly the wind drops. The elders know that the center of the storm is overhead. They tell everyone to move inland without delay, for they know what approaches from the ocean. Men, women, and children gather their belongings and run for such higher ground as there is in the flat landscape. The wind returns without warning, stronger than before and with it the ominous sound of rushing water. Within minutes, the villagers find themselves lying on a low island that was once a ridge. Blowing spray soaks them to the skin as they clutch tree trunks and hold on for their lives. Two women lose their grip and vanish into the gloom, never to be seen again.

Hours later the wind drops. Nothing remains of the village except a few floating timbers and carcasses of water buffalo drowned by the surge despite being good swimmers. Seawater slowly recedes from the waterlogged countryside, leaving a muddy wilderness of shattered trees and soggy grass in its train. Dense clouds of steam rise from the sodden delta as the hot sun and suffocating humidity torment the land. The stink of decaying bodies and rotting vegetation pervades the air. The farmers pick themselves up. They slowly rebuild their village and replant their rice gardens with the arrival of the monsoon floods. Severe typhoons are part and parcel of human existence along the East China Sea, and have been since the time of the ancestors.

LIFE ALONG CHINA'S long, low-lying coastline has involved constant adjustment to changing sea levels ever since the end of the Ice Age. Many people call this a "muddy coast," a landscape of mudflats and shallow lakes, of wetlands and sand dunes.[1] As is the case along all "muddy" coasts with reduced topography, sea level rise can cause extensive horizontal water movement, which is often out of proportion to the actual extent of the rise. Waterlogged environments, often rich in game and plant foods, expand and contract with even minor changes in the ocean. When people lived off game, fish, and plant foods, adapting to these shifts was a simple matter, even when hunting groups lived in the same location for long periods of time. Relationships with the changing sea became more complex when coastal peoples began cultivating rice, today the staff of life for the Chinese.

Rice accounts for over half the food eaten by nearly two billion people and over 21 percent of all calories consumed by humankind. Yet the history of domesticated rice remains largely a mystery, except for the certainty that the Lower Yangtze Valley played a critical role in its early cultivation.[2] Unfortunately we have little to go on, except for some charred rice fragments from the walls of a pot found in the delta region dating to around 8000 B.C.E. Much farther upstream, in the Middle Yangtze region, minute rice phytoliths, microscopic fragments of intercellular silica, appear in occupation levels at Diaotonghuan Cave, dating to around 10,000 B.C.E.[3]

Both these finds were wild forms, for it took five thousand years for fully domesticated rice to develop and become genetically "fixed." Many generations of early cultivators grew some partially domesticated rice, while still relying on many species of other wild plants, among them

Figure 9.1 *Map showing locations in chapter 9.*

fruit and acorns. The crops they planted interbred with wild rice stands, resulting in considerable diversity due to a high rate of genetic exchange. It was not until the nascent farmers created isolated rice plots that fully domesticated forms developed and became fixed.

Much of this cultivation took place in wetland environments, such as those of the Lower Yangtze delta, where wild rice, adapted to marsh environments, once abounded. Such rice appears to have spread into the

Yangtze valley around fourteen thousand to thirteen thousand years ago, when climatic warming brought trees and other new plant forms northward. This indigenous rice is long extinct, but some of its traits passed down the generations to become part of domesticated forms.

The low-lying coastline with its lakes and marshes may have been an ideal environment for rice cultivation, but the farmers had to cope with ever-changing sea levels, especially around 5000 to 4000 B.C.E., when sea levels were several meters above modern levels, just before glacial melting slowed. Lake Taihu, south of the Lower Yangtze River, a region of marshes, ponds, and larger bodies of water, was a magnet for hunter and rice farmer alike in 5000 B.C.E.

Excavations at a waterlogged site near Kuahuqiao, some two hundred kilometers southwest of Shanghai, tell us of struggles with an encroaching ocean at a time when warmer conditions and a strengthening summer monsoon brought higher rainfall and more severe flooding after 7000 B.C.E.[4] The village would have been near invisible among the grass and reeds of the coastal wetlands, a cluster of thatched dwellings set on piles above the swampy, often waterlogged ground. Only dogs barking, the sounds of children at play, the pounding of mortars, and the scent of woodsmoke would have told of human habitation. What is now called Kuahuqiao lay at a strategic point between upland valleys and coastal wetlands along Hangzhou Bay. The first settlers arrived around 6000 B.C.E., settling among swamps at a time when today's lowland terrain was not fully formed and sea levels were rising, not only in the East China Sea but also globally. They cleared the scrub ground cover and planted small amounts of rice. Apparently the planting was a success, for rice pollen grains soon become common in the surviving occupation levels. So do the ova of a parasite worm that is associated with both humans and pigs.

After about a century, rising water flooded the site. So the inhabitants moved, something that was relatively easy for them to do. Local populations were thin on the ground; there was plenty of land above water level and abundant land for crops, humans, and animals. The water apparently receded, for later generations of rice farmers moved back to Kuahuqiao. The excavators believe that the inhabitants grew rice in fields that were flooded regularly, then burned off and manured with animal dung to enhance crop yields. They turned the soil with shovel-like artifacts made from pig shoulder blades. We know from the excavations that cattails grew thickly around the cultivated fields, themselves an important food source and useful raw material for a variety of purposes. Not that these were full-time rice farmers, for they also relied on wild plants and hunting, while dogs and pigs roamed their settlement.

Generations of farmers occupied the same general location for as long as four hundred years. Judging from the thicket of house piles underfoot, the village was rebuilt many times until about 5550 B.C.E., when marine flooding apparently caused by rising sea levels made cultivation impossible.

Even before then, the farmers had to handle encroaching saltwater, but they apparently did so with some success, for the earliest rice culti-

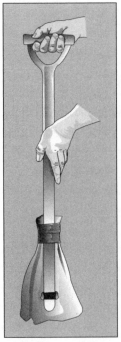

Figure 9.2 *A shovel from Hemudu with a blade made from a water buffalo scapula. Author's collection.*

vation in eastern China took hold in slightly brackish coastal reed
swamps. From the earliest days of agriculture, the new farmers cleared
much indigenous vegetation and managed the coastal marsh environment with fire. At the same time, they appear to have taken careful steps
to control the amount of tidal seawater that reached their fields. They
would have used artificial dikes to retain nutrient-rich freshwater, to
prevent catastrophic flooding, and to provide the consistent water regime that rice requires.

Kuahuqiao is by no means unique, but the rice-farming population
in the Lake Taihu region was never large, concentrated as it was in wetland and swamp areas along the humid Lower Yangtze and on either
side of Hangzhou Bay to the south of modern-day Shanghai. We have
but occasional snapshots of these farmers, one from Hemudu, a village
in the same general location but at a lower elevation where silt deposition from river floods and sea level changes exposed more land. Like
Kuahuqiao, Hemudu lay in an ideal environment for cultivating rice,
especially after 5000 B.C.E., when wild nut-bearing trees declined in the
face of warmer and wetter conditions and people may have turned to rice
as their staple.[5] Hemudu covers an area of about four hectares, occupied
on at least four occasions, starting as early as 5000 B.C.E. The main occupation ended some four hundred years later, around 4600 B.C.E. The village lay in a forested landscape surrounded by ponds, where the people
apparently processed rice on a large scale, but whether it was fully domesticated is a matter for continuing debate. Certainly rice was an important component in daily life, grown in waterlogged landscapes where
excellent preservation conditions have yielded beautifully made mortice-
and-tenon-joined building planks and hoes with blades made from water
buffalo shoulder blades.

Small villages like Hemudu nestled among wetlands, where human
occupation ebbed and flowed with rising sea levels, changing flood conditions, and sediment buildup. Over thousands of years, small-scale agriculture combined with foraging and hunting slowly gave way to more
intensive rice cultivation with fully domesticated forms. Population
densities remained low in the Lower Yangtze until around the later fifth
millennium B.C.E., when they rose rapidly. Domesticated rice grown in

wetlands is highly productive; inevitably farming populations now exploded and the delta landscape became more crowded. Much larger, more permanent settlements flourished, most of them lying close to extensive rice paddies, protected from sea surges by coastal marshes. As populations rose and villages grew larger, so the waters of the Yangtze became a greater threat than rising sea levels.

THE YANGTZE (sometimes called the Cháng Jiang) is the longest river in Asia and the third longest in the world. It rises on the Tibetan Plateau, then flows 6,418 kilometers across southwestern, central, and eastern China before discharging through an extensive delta into the East China Sea near Shanghai. Dozens of tributaries and smaller streams, and also numerous lakes, contribute water to the Yangtze as it drops from an altitude of about 4,900 meters to sea level. Before the building of the Three Gorges Dam in Hubei Province, oceangoing vessels could sail as far as 1,609 kilometers along what is sometimes called the Golden Waterway. Like a hydrological octopus, the Yangtze and its many tributaries and nearby lakes affect the rise and fall of sea levels thousands of kilometers from its source. The story of sea level changes and their effects on those who live in the Yangtze delta involves not only changes in coastal geography, but also the vagaries of a river plagued with monsoon-generated floods.

Like the Nile, the Yangtze bears a huge silt load downstream, much of it ending up at the coast. Some 170 million cubic meters of silt once flowed downstream to the coastal plain annually—before the construction of the Three Gorges Dam, which reduced the sediment load significantly. The river lies on a rainfall frontier, created by the north-moving summer monsoon. Regions south of the Yangtze receive their rainfall in May and June. To the north, rains come in July and August. Inevitably, a strong monsoon leads to heavy flooding and disaster, usually between May and August. The river overflows its banks and inundates thousands of hectares of the surrounding landscape. Villages are swept away, thousands are drowned, and crops are decimated. The casualties from twentieth-century Yangtze floods alone have killed over 300,000 people.

In 1931, 40,000 people perished in a major flood; 30,000 drowned in a 1954 inundation and an additional 200,000 died from starvation and disease. A villager who lived through the 1954 disaster told a reporter from the *Washington Post* that "the corpses were put in coffins but they could not be buried. They were just stacked up."[6] The year 1998 brought the worst floods in forty-four years, which killed more than 4,000 people and left 13.8 million homeless. A total of 240 million Chinese, a figure equivalent to the population of the entire United States, were affected directly by rising water.

Over the past thousand years, the intervals between major floods have shortened.[7] Between 618 and 907 C.E., major inundations occurred about every eighteen years, increasing to every four to five years during the Song and Qing Dynasties (960 to 1911). Since 1950, the recurrence has been about every three years and perhaps changed during the past quarter century to about every four. Data from flood gauges recorded since the nineteenth century shows that the peak water level has also increased, from about 27.4 meters in 1870 to about 30 meters in 1954. Now the levels are even higher. At the same time, the duration of the peak water level has increased from about six hours in 1973 to just over three days in 1998. More flooding may also be occurring in the Yangtze delta owing to more frequent outbreaks of heavy rainfall caused by global warming.

For thousands of years, farmers living along the Yangtze's banks adjusted fairly readily to even severe inundations, for the numerous lakes on either side of the river handled much of the overflow. However, the rapid population growth and industrialization of the past century have raised the stakes, especially as sea level rises now present a serious threat to cities in the Yangtze delta.

The history of Chinese sea level fluctuations is hard to decipher. We know that sea levels retreated somewhat after 4000 B.C.E. as a result of local crustal adjustments, additional glacial meltwater in the ocean that tilted continental shelves upward, and also water loading, a global phenomenon described in chapter 1. Then the coastline tended to stabilize for several thousand years, during the apogee of Chinese imperial civilization, the level changing perhaps no more than a quarter of a centimeter a year.

For all its catastrophic river floods, the Yangtze became a breadbasket of rice growing during the more stable sea level centuries, and assumed great importance in later Chinese history. Over the centuries after the abandonment of Hemudu, population densities rose, villages grew larger, and towns came into being where artisans and traders flourished. Specialist workers began to exploit the mineral wealth of the Yangtze valley—copper, tin, lead, and salt.

Of all these commodities, salt was in many respects the most vital, for it was an essential to subsistence farmers relying on predominantly carbohydrate diets.[8] It was also valuable for hide processing. The major salt deposits lay to the north, which may have led to contacts with communities as far away as the Huang He River valley. Salt was to become a major trade commodity in later times, especially with the rise of the Shang civilization in the Huang He River valley after 2000 B.C.E. Copper abounded in the Middle and Lower Yangtze region, while lead and tin came from deposits south of the river. All of this activity resulted in growing contacts between north and south, especially with the emergence of intensely competitive states in the Yellow River valley, where copper and bronze technology reached a high level of sophistication.

The farmers of the Yangtze delta thrived off long-distance trade. Between about 3400 and 2250 B.C.E., the Liangzhu culture prospered in the region, still a farming society based on paddy farming of two rice species, and still a society where most people lived in silt houses along rivers or along shorelines, which must have given them at least some protection against floods and sea surges. By now, however, local society was much more sophisticated. Liangzhu became a small state presided over by a wealthy elite. We know their leaders from their burials, which contain elaborately incised ceremonial jades, symbols of military and ritual power. The jade trade was an important part of this society, whose lords wore silk garments and prized ivory and lacquer artifacts.

Liangzhu extended from Lake Taihu south of the Yangtze as far north as Nanjing and into the present-day Shanghai region. Its cultural influence spread over an enormous area, as far north as the Shang civilization capital at Anyang near the Yellow River. Much of the major activity lay inland, away from the delta lowlands, but it is clear that the adoption of

more sophisticated, high-yielding rice cultivation provided the base for
a much more elaborate farming society, but one where many people
were still forced to move as sea levels fluctuated and in response to major river floods. By this time, however, sea levels had more or less stabilized to modern levels, which made the farmers' task somewhat easier
than in earlier centuries.

Over hundreds of years, the Yangtze Valley became a cockpit of competing states and chiefdoms until the first emperor, Qin Shi Huangdi,
unified China in 221 B.C.E. The succeeding Han dynasty, which flourished from 206 B.C.E. to 220 C.E., brought great economic prosperity and
fostered stable, highly productive irrigation agriculture, which endured
into historic times. A wealthy and sophisticated imperial realm developed, to be sure, but one plagued by severe floods and droughts resulting
from monsoon failures that killed tens of thousands.

LIFE ALONG THE Lower Yangtze was a constant adjustment to the realities of life very close to sea level. In time, high levees protected much of the
surrounding countryside from the flooding river. But when these levees
broke, water swept across the surrounding landscape, bringing death and
disease in its train. The delta was home to small communities of rice farmers, who had more problems with river floods than with sea levels during
the imperial centuries. Their difficulties with the ocean came from sea
surges generated by typhoons, but these were temporary episodes, expensive in terms of destruction and human life without question, but not a
permanent encroachment on land they had farmed for many generations.

There were some towns and small cities, notable among them a market
town, now named Shanghai, founded as early as the tenth century C.E. in
a swampy area east of Suzhou in the delta.[9] Shanghai prospered modestly
before becoming a major center of cotton production and textile milling
of all kinds. Silk from the Suzhou region became famous throughout
China, creating a class of wealthy estate owners. These industries remained the mainstay of the Shanghai economy until the nineteenth
century. The First Opium War (1839–1842) between Britain and China

ended with the signing of the Treaty of Nanking, which created five treaty ports for international trade. One of them was Shanghai, whose close proximity to the mouth of the Yangtze made it an ideal location for an explosion of trading activity.

Shanghai soon became a boom city. By the 1860s, the city boasted of a significant population of British and American merchants, who formed an international settlement centered on the waterfront area still known as the Bund. Three quarters of a century later, three million people lived in the city, now one of the largest in the world. Thirty-five thousand foreigners, Shanghailanders, controlled half of Shanghai, now the commercial center of Asia. The Japanese occupation of World War II brought the boom to a halt. Nor did the city prosper under early Communist rule, when it bore a heavy tax burden. Everything changed during the 1990s, when Shanghai-born politicians achieved some dominance in the central government. Since then, the city has been a commercial powerhouse, handling over a quarter of the trade passing through Chinese ports, an economic boom with potentially disastrous environmental consequences.[10]

Figure 9.3 *Shanghai's Bund in about 1935.* © *Albert Harlingue/ Roger-Viollet/The Image Works.*

Today, over twenty-three million people live in the city and its environs, the population growth caused not by the birth rate, which is lower than that of New York, but by continual in-migration from people seeking work and opportunity. The administrative area of the city includes significant agricultural land, but construction proceeds at an astounding level, much of it fueled by government investment. Population growth, exploding industrial activity, and many more motor vehicles have created serious and in some respects unexpected environmental problems that not only place the low-lying city at serious risk from water pollution and solid-waste issues, but also, most important, from sea level rise.

Shanghai lies on the East China coastal plain, the city and its environs located on an alluvial flatland that forms a bulge of the Yangtze delta, averaging about 4 meters above the low tide mark. Since high tides can occasionally reach 5.5 meters, the city already relies on dikes and embankments to prevent flooding—this before one factors in rising sea levels. Until the construction of the Three Gorges Dam far upstream, the Yangtze generated a huge runoff of freshwater each flood season, delivering about 486 million metric tons of silt to its estuaries every year. The sediment formed submerged deltas and tidelands that gradually formed land ever farther out to sea, despite rising sea levels.[11]

Like deltas everywhere, that of the Yangtze is subsiding every year. Until recently, the subsidence has always been due to natural tectonic forces. From 1921 through 1948, land subsidence in Shanghai was 2.4 centimeters a year. As industrial activity picked up after 1949, groundwater extraction accelerated dramatically. The underlying sediment became compacted and subsidence increased rapidly. By 1965, the cumulative effects formed two large depressions that covered more than four hundred square kilometers, the subsidence level reaching 2.63 meters, despite efforts to reduce groundwater pumping. At this point, fifty square kilometers of the city along the Huangpu River were below normal high tide level. Despite artificial groundwater replenishment, reduced pumping, and adjusted pumping layers, measures that are said to have brought the subsidence under control, the land surface continues to sink at a rate of 4 millimeters annually in the city, even more in suburban industrial districts, where pumping still exceeds replenishment by 30 percent or more.[12]

Ground subsidence obviously contributes to Shanghai's vulnerability to rising sea levels. Flood defenses along the Huangpu River and in Suzhou Creek, major port areas, have already been raised several times, but they are still inadequate protection. The construction of floodgates and raising the existing flood walls even farther will require enormous financial investment on the part of the city.

HERE, AS ELSEWHERE in the world, significant sea level rise resumed after the mid-nineteenth century. According to official Chinese records, sea level rise along the East China Sea coast has averaged between 14 and 19 centimeters over the past century. A rise of about 2 to 3 millimeters a year is predicted for the foreseeable future. After correction for vertical movements caused by subsidence and earth movements, the average rise is about 2 millimeters annually. The Intergovernmental Panel on Climate Change (IPCC) has produced estimates of 18 centimeters by 2030 and 66 centimeters in 2100. The highest estimate is 110 centimeters. All these projections assume what is called business as usual—no efforts to curb greenhouse gases. However, it is hard to estimate sea level trends in the Yangtze delta, because of river runoff, the constant shifts in river channels, and land subsidence caused by pumping out groundwater. Expert calculations using tidal gauges and other data are in the 2.6 centimeter range. As far as one can discern, the rate of sea level rise in the delta is somewhat higher than elsewhere along the coast, perhaps because of river runoff, but everywhere, the trend is toward ever-higher levels.

Sea level rise in the Shanghai city area itself during the twentieth century has exceeded two meters, which represents a significant threat to the city and surrounding coastal plains. Now that the Three Gorges Dam on the Middle Yangtze has significantly reduced sediment discharge from upstream, the menace of accelerated coastal erosion also raises its ugly head.[13] (Fortunately, tributary rivers still bring down large sediment loads.) River runoff and shifting river channels are the major source of erosion, for they cut back the low-lying coast and narrow, or completely remove, tidal flats that are valuable coastal protection. Nearly half the coastline in the Shanghai municipality is described as "eroding," so much depends on the silt loads that arrive from upstream. If there is not enough sediment in future decades, rising water levels will decimate tidal flats and intensify erosion throughout the estuary. Widespread land reclamation and clearance of mangroves and marshes for agriculture and development add to the threat.

Subsidence, in part due to headlong development, especially the construction of high-rise buildings, means that much of Shanghai is only 1.5 meters above low tide level. In 1999, existing dikes and flood walls protecting the city extended 465 kilometers, with an average height of eight to nine meters, designed to withstand a Force 10 typhoon and a fifty-year high tide. They are currently being upgraded to withstand a thousand-year flood, but standards in the surrounding agricultural regions are still poor. Without such defenses, almost all of the city and surrounding districts would be flooded with every high tide. A half-meter rise in sea levels would inundate 855 square kilometers of the city, port, and environs. Nearly all the city would disappear underwater if the level rose by a meter. At the same time, sea level rise would aggravate the salinization of the soil and increase chronic waterlogging of low-lying agricultural land. The Pudong area with its major industrial facilities would lie below the water with a sea level rise of a half meter or slightly more. The only response short of the impossible task of moving the entire city is to raise and consolidate the existing coastal dikes and flood walls to prevent catastrophic inundation.

Then there are the effects of sea surges generated by typhoons and major gales. About two typhoons hit Shanghai each year; the highest sea surge on record reached 5.22 meters near the city center in 1981. A combination of sea level rises and a predicted higher frequency of typhoons in the future may, theoretically, have the effect of reducing the thousand-year flood frequency to a century or even less. Furthermore, heavy rains after typhoons and other major storms result in high flood levels upstream, which cause extensive waterlogging in the delta area.

High sea levels would also reduce the amount of runoff that can be handled naturally by gravity, so expensive pumping may be necessary.

Overpumping groundwater to supply cities and nourish irrigated fields as well as promiscuous construction in coastal cities like Shanghai has exacerbated the process, causing sediments to compact and sea levels to rise sharply. The faster-rising ocean of the immediate future will deepen the continental shelf close offshore and make it harder for waves to disperse coastal sediment. River gradients will become shallower, as they have elsewhere, decreasing sediment discharge. So will human activity upstream, like the construction of the Three Gorges Dam, which has reduced sediment discharge even further. Already coastal sand barriers are retreating; sediment is reduced, so beach erosion increases, especially with prolonged La Niña conditions and a predicted higher incidence of severe storms and their accompanying sea surges. As the sea level rises, so the coastline retreats, triggering significant environmental changes farther inland. Saltwater intrusion intensifies and freshwater supplies for towns and villages are affected, especially during the dry season. Between December and March, when water levels are lower, the authorities already restrict water supplies from the heavily polluted Huangpu River that flows through the city. Future sea level rises may seriously affect water quality in Shanghai at a time when local aquifers are already well drawn down, but fortunately not yet affected by seawater. The government is taking urgent steps to replenish groundwater not only in the Shanghai region but also elsewhere, but this will be effective only if strict controls govern the quality of the replenishment supplies, something that has been erratic in the past.

Almost half of Shanghai's coastal zones have been reclaimed from marshes and tidal flats, this for an area that produces about a ninth of all China's agricultural and industrial output, with but 1 percent of the national population. Shanghai is the most densely occupied city in China, but most of its more than twenty million already live below the high tide mark—or would do so if the ocean rose by a half meter. The situation is somewhat akin to that in Bangladesh, with the long-term prospect of millions of people becoming environmental refugees. The cost of resettlement would be almost unimaginable, leaving effectively only one option—huge capital expenditure on flood defenses. The future depends on effective long-term planning of all construction in areas subject to erosion and potential inundation, strengthening of seawalls, and systematic experiments with salt-resistant crops while accelerating land reclamation from marshes and tidal flats to tap the potential of foreshore areas as quickly as possible. A range of other measures might also be effective. One option might be to construct freshwater reservoirs upstream to receive floodwaters, which could then be released during the dry season. But here again there are environmental consequences, for a reduced flood downstream would also mean less sediment deposition, which would not be so bountiful if the flow came from controlled reservoir releases.

All of this requires closely integrated, even authoritarian, planning and proactive construction of sea defenses and other facilities against the day when sea levels will be effectively higher than Shanghai. An expensive era of very long-term planning and massive capital expenditure for a future beyond the lifetimes of those currently living and working

Figure 9.4 *Water lock employees examine the water lock at Suzhou, Shanghai, in 2009, about three meters above sea level, designed to protect the city against sea level rise. AP Photo/Eugene Hoshiko.*

in Shanghai may yield few immediate benefits in today's world. But future generations will bless the foresight that protected one of the great cities of the world from inundation.

Shanghai authorities think that the city can survive rising sea levels until 2100 with currently enacted or planned infrastructure investments. But longer-term planning requires different measures. One palliative under consideration involves the construction of a floodgate near the Yangtze estuary, the gates being raised or lowered according to the dictates of tide and weather, thereby controlling water flowing in and out of the river. Shanghai has already spent over six billion dollars on flood defenses over the past decade; a floodgate might be more cost-effective, despite concerns over the environmental and commercial impacts of the project. Meanwhile, the city is becoming ever more vulnerable, in part because of aggressive reclamation of coastal wetlands in response to a land shortage. These wetlands have provided a natural barrier against inundation for hundreds of years. Now most of them are gone. In the long term, perhaps the best strategy might be to move away from expanded industrial activity into less aggressive impacts on the environment. For instance, the city is planning to foster economic activities such as banking, which do not require land-hungry factories.

WE HUMANS ARE opportunistic and voracious when we settle at strategic locations on the world's coast. Shanghai epitomizes the dilemma faced by an industrial economy that depends on foreign investment and trade for its livelihood. Protective wetlands vanish daily; aggressive industrial development continues unabated. The authorities invest heavily on sea defenses, with no guarantee that they will work in the very long term. They really have no other option, given the impossibility of moving millions of people inland. Shanghai faces the same unhappy prospects as Bangladesh—millions of people, settled on low-lying terrain, powerless to react to an ever-accelerating vulnerability. The future will define Shanghai's appalling dilemma with frightening clarity. Solutions lie not in immediate panic measures or inflated political rhetoric, but in deliberate, carefully measured long-term planning.

"Wave in the Harbor"

NORTHERN JAPAN, A SUMMER MORNING, 3000 B.C.E. Clusters of brush dwellings dug partially into the ground lie in a forest clearing atop the ridges overlooking the bay. Woodsmoke wafts idly above the dark tree-tops. Two women scrape a hide pegged out on the ground near a smoldering hearth. A young man and his father emerge from the trees hefting the carcass of a deer. Children cluster around the fresh kill asking questions about the hunt. The village looks down on the wide bay and its sandy beach, where dark figures climb on the rocks, fish spears in hand. Three young girls pry mollusks off rocks exposed by the falling tide and cast them into a skin cloak that doubles as a container.

High above on the ridge, three elders gaze out over the ocean, as they do every morning as the sun climbs in the heavens, talking quietly among themselves. Suddenly the ground starts shaking. Startled birds fly from the trees in cacophonous fear. The heavy shaking continues. An irregular rolling and thumping dislodges boulders from the nearby cliffs. Except for the children, everyone has experienced this before. Both in camp and by the shore, everyone crouches instinctively on the ground, mothers clasping their children. Fishermen on the rocks jump for safety into the sea.

As the shaking subsides, the elders leap to their feet and watch the ocean, calling loudly to everyone and beckoning people by the shore urgently back to camp and onto higher ground. Suddenly the water starts to recede far below low tide. As fish flap helplessly and the ocean becomes dry land, young and old alike run as fast as they can up the ridge and then to even higher ground behind the camp. Safe above the bay, the people turn seaward just as some loud booms and a deep roar assault their ears. A great wave flows rapidly inshore and breaks loudly at the foot of the ridge, throwing spray high into the air. Moments later, an enormous wave follows in its wake, this one far higher than two or three men. Great torrents of water surge up the slopes of the ridge, slowing just as they reach more level ground and the edge of the village. More waves attack the shore, but with less force than their predecessors. Soon the water retreats as fast as it arrived and the ocean resumes its normal calm. The beach and the low ground behind it are now a barren landscape of sand and rocks, but, thanks to deeply ingrained cultural awareness of tsunamis passed from one generation to the next, the villagers have survived.

AN EARLY ACCOUNT of a Japanese tsunami appears in the *Nihon San-dai Jitsuroku, The True History of Three Reigns of Japan*, written in 901 C.E. On July 13, 869, "Some time after severe seismic shocks, a gigantic wave reached the coast and invaded [the] entire Sendai plain. Rising seawater flooded an old castle town . . . causing the loss of 1,000 lives."[1] The word *tsu-nami*, in use in Japan since the 1600s, means "wave in the harbor," apparently a reflection of the devastation witnessed by fisherman returning home after a major tsunami had swept ashore. Great tsunamis punctuate Japanese history from its beginning. We can be certain that vivid recollections of many of them remained etched into generational memory for centuries afterward. Tsunamis offer a dramatic lesson in the importance of cultural traditions in surviving the vagaries of the attacking sea.

Japan has always rocked and rolled, lying as it does on the notorious Pacific Ring of Fire. Around the Pacific Rim, great tectonic plates clash against one another, displace seawater, and give birth to destructive tsunamis. Some 53 percent of all tsunamis occur in the Pacific. Over 80 percent of them result from submarine earthquakes. Tsunami waves move forward with tremendous power, literally a solid wall that acts like a bulldozer. Instead of expending its force by breaking, a tsunami continues to surge inland until friction or a steep gradient slows it. Once the momentum of return develops, it is extremely powerful and drains away with tremendous force, carrying people and property, even entire buildings, out to sea.[2]

As the world learned the hard way from the Aceh tsunami that struck Southeast Asia and the Indian Ocean in 2004, such events are about the most dangerous assaults the ocean can unleash on humanity. They strike without notice, give virtually no hint of their coming, and can wipe out virtually everything in their path. You can forecast a tropical cyclone and its sea surge a few days ahead, enough time to track their arrival. But tsunamis develop, sometimes without warning, are usually lethal, and, in a future world of higher sea levels and much denser coastal populations, are a frightening menace.

TSUNAMIS ATTACKED JAPAN long before humans settled there some twenty-five thousand years ago, but it was not until thousands of years later that they caused major social disruption. During the late Ice Age, Japan's four main islands formed a single landmass, connected to the Asian mainland. A land bridge joined what is now Hokkaido in the north to Sakhalin. Coniferous forests covered the landscape at a time when temperatures were much lower than today across a tract of dry land that was about the size of California. When global warming accelerated after fifteen thousand years ago, sea levels rose quickly in Japan, just as they did elsewhere, at times as much as 2.5 meters a century, owing to rapid release of glacial meltwater. The overall sea level height was about twenty to thirty meters below modern levels eleven thousand years ago. Thereafter, the climate warmed rapidly. Sea levels reached a height about two to six meters above modern levels around 5400 to 3900 B.C.E., just as they did along the coast of the East China Sea. Thereafter there was a slight retreat to about one to three meters below modern levels between 2500 B.C.E. to 1 C.E. in some places.

Profound environmental changes accompanied sea level rise. The coniferous forests of earlier times gave way to temperate deciduous and evergreen broadleaf forests. Nut trees abounded. What had once been

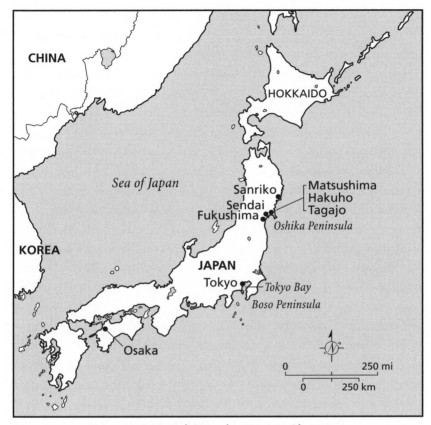

Figure 10.1 *Map showing locations in Chapter 10.*

a plain became the shallow Sea of Japan. The new archipelago abounded in sheltered bays, tidal flats and wetlands, and estuaries where fish, mollusks, and other seafood abounded. At the same time, the large game animals such as the elephant and bison of earlier times gave way to medium-size and small species such as the sika deer and wild pigs.

The newly fashioned archipelago was a veritable paradise for the hunters and foragers who had eked out a living in what is now Japan for at least twenty-five thousand years. As the vegetation changed and sea levels rose, Jomon society emerged on the islands—hunters, foragers, and fisherfolk, who may have originated on the Asian mainland, perhaps in the Amur River valley of Northern China.[3] The archaeological term Jomon means "cord-decorated," after the distinctive cord-decorated pots found in their settlements, for the Jomon people were among the earliest potters in the world—at least sixteen thousand years ago.

Bark, leather bags made from stomachs and intestines, ostrich egg-shells, seashells, and wooden trays—ancient hunter-gatherers used an impressive array of simple containers, many of them improvised for specific tasks like collecting honey. Bowls and pots fashioned from fired clay are quite another matter, not because someone invented a brand-new technology—fired clay was familiar to Ice Age people in many places as early as twenty-five thousand years ago. What was new were durable, waterproof receptacles, which could be used to store food and water, and, above all, for cooking or for steaming various foods. Shellfish could be opened easier; small children could be fed soft foods, thereby shortening weaning periods and permitting closer-spaced births; toothless old people could now be fed easily, allowing them to live longer. Since the elderly were repositories of ritual and spiritual belief, of cultural knowledge of all kinds, their potentially longer life expectancy was no small matter.

This would have been especially true in a world where many people lived by the coast and were at risk from tsunamis. Thanks to an abun-dance of nuts and other plant foods, many groups stayed in the same villages for long periods of time. Reduced mobility increased vulnera-bility to tsunamis, making generational memory of such events even more important. So did an increasing reliance on estuaries and bays, on coastal wetlands, fish, mollusks, and sea mammals, foods so abun-dant that people lived for long periods in the same spots, close to the high tide mark. The ocean, like the land, became a conscious part of the Jomon landscape.

Imagine a calm bay, deep water lying close offshore. A Jomon village lies on a low headland overlooking the ocean. On calm days, the hunters watch dolphin gamboling close inshore. An unsuspecting school feeds in the deeper water of the continental shelf close to the inlet. The fisher-men launch their dugout canoes and paddle a short distance offshore, piles of stone cobbles in the bottoms of their craft. They fan out to en-circle the feeding mammals. The leader gives a signal. The crewmen grab cobbles and knock them together below the surface. Confused by the unfamiliar cacophony, the school swims to and fro, disoriented by the unfamiliar resonance. The lead dolphin steers the confused beasts away from the noise into the shallows of the narrow bay as the canoes close in. Men, women, and children rush into the shallows and literally fling the helpless dolphins ashore. Sometimes they grab the beasts, holding them softly by the mouth and guiding them alongside the canoes where spearmen dispatch them in short order. Once the hunt is over, the butchery begins. The women cut long strips of dolphin flesh and hang them on wooden racks to dry in the sun and wind.

It is an imaginary hunt, but a scenario based on traditional dolphin hunting methods that were commonly used in many parts of the world in ancient times. Jomon groups did far more than hunt dolphin. Entire villages harvested the great salmon runs of spring and fall, used river weirs to trap smaller fish, often prized not so much for their flesh as their high oil content. Nor were Jomon fishermen afraid to venture into deeper water, where they caught tuna as they migrated close inshore. Many of the species they took, like salmon or tuna, yielded flesh that could be smoked or dried for later use, providing a stable food supply so that many coastal settlements remained in use for many months, even year-round.[4]

Quite apart from staples like dolphin and salmon, every Jomon village close to shore clubbed seals when rookeries were nearby. The entire coastal landscape created by rising sea levels was an ancient supermarket of different foods—crabs and oysters, and shellfish of all kinds, many of them taken by diving into deeper water. Some Jomon skeletons display the characteristic abnormal growth in the ears found in many modern divers. It's no coincidence that huge shell mounds accumulated around Jomon villages in places where mollusks abounded. They were an invaluable staple, especially when fish or plant foods were in short supply during the winter months.

Anyone living by the shore learned about the ocean landscape literally from birth. They acquired knowledge from their parents, from elders and fellow kin, and from hard-won experience on the water and on shore. If the Jomon were like other hunting societies, they would have lived in an intensely symbolic world, where ceremonies and chants, dances, myths, oral traditions, and songs, both connected them with the spiritual world and defined the cosmos, its benefits, dangers, and hazards. And part of this complex ritual tapestry would have been memories of great tsunamis and powerful earthquakes, carefully recounted so as to prepare everyone for sudden flight to higher ground when an earthquake struck. Such behavior would have been as much a part of daily existence as preparing a meal or harvesting nuts.

Just as they did in Europe, rising sea levels nurtured rich inshore fisheries, while post–Ice Age warming created much richer environments onshore. Most Jomon communities relied heavily on plant foods, especially the rich fall nut harvests, which yielded millions of chestnuts and walnuts, and also horse chestnuts and acorns. Nuts have huge advantages; not only are they nutritious, but they can also be stored for months, even years, in pits or granaries. Jomon foragers stored the harvests in subterranean pits up to 1.8 meters deep. With such diverse and plentiful food sources at hand, population densities rose steadily, especially in parts of food-rich eastern Japan, so much so that the most favored areas became crowded and hunting territories more circumscribed. There may have been food shortages, too, especially of easily stored nuts. In response, Jomon communities made increasing use of carbohydrate-rich acorns, which require labor-intensive shelling, then grinding before processing them to remove the bitter tannic acid, which has to be leached away by immersing the ground nuts in water or by flushing water through a pile of nuts in a hollow.[5]

No one knows how large the population was after thousands of years of successful fishing and foraging, but one conservative estimate speaks of some 250,000 Jomon people throughout the islands. Their villages were certainly thick on the ground, especially in food-rich areas. Some of them boasted of more than fifty houses, many located close to water's edge. Population densities rose so high in some regions like Tokyo Bay that individual territories for each settlement may have been as small as five kilometers across. A combination of higher numbers and more permanent settlement increased Jomon vulnerability to tsunamis, but it was a vulnerability mitigated by strongly centered cultural traditions that stressed movement in the face of disaster. There may have been tsunamis aplenty during the Jomon millennia, but casualties must have been relatively small when everyone knew instinctively what to do when a strong earthquake shook coastal villages.

Local populations waxed and waned over the centuries, but increasingly complex Jomon societies flourished for more than ten thousand years. They thrived through thousands of years of often abruptly changing environmental conditions and warming, and into the period of more stable sea levels, which took hold all over the world around 4000 B.C.E. They adapted to rising sea levels—as did hunter-gatherers elsewhere. A remarkably high level of cultural continuity, and presumably a high degree of adaptiveness to tsunamis, endured over the islands until as late as 300 C.E., when new rice-farming societies, known collectively as Yayoi, came into being in the south.

One uses the words "came into being" advisedly, for controversy reigns over exactly what happened. For many centuries there had been at least sporadic contacts between Jomon communities and Korea, some two hundred kilometers away, but it was not until about 400 B.C.E. that these contacts intensified with the arrival of rice paddy cultivation, domesticated pigs, and full-time farming in Japan. On the southwesternmost island, Kyushu, the farmers may have grown rice in the paddies during the warm summers, then drained them for dry cultivation of cereals like millet and wheat in the winter. This highly productive form of intensive agriculture triggered rapid population growth as farming spread rapidly northward as far as Honshu within three hundred years. Whether this was a case of Jomon people adopting a totally new way of life, or the arrival of significant numbers of migrants from Korea, where similar artifacts to those of Yayoi farmers are found, is a much-debated and unresolved question. In the much cooler far north, rice farmers had no hope of competing with hunters and gatherers. Northern Honshu was the frontier between farming and the ancient lifeway, which survived in the hands of the Ainu people, genetic descendants of the ancient Jomon, until modern times.

WHEN THE GREAT tsunami of March 11, 2011, thundered ashore along the coastline around Sendai in northern Japan, huge waves swept away everything in their path. But Jomon sites dating to around 4500 B.C.E. on the hills and terraces above the shore at Satohama in Matsushima Bay remained unscathed.[6] These extensive sites lie some twenty to thirty meters above sea level, out of reach of almost all tsunami waves. Later Jomon sites lie at slightly lower elevations, while artifacts left by rice farmers from the ninth to twelfth centuries C.E. lie around today's high water mark. Were the inhabitants of Satohama in the ancient times aware of tsunamis? We can be sure that the Jomon had experienced them, which may be why they built some of their villages well above sea level. The pattern of settlement changed with Yayoi rice-paddy farmers, who required low-lying, easily inundated land, such as lay on coastal plains. Higher population densities, more people packed into the landscape, all of them anchored closely to their fields and paddies: The threshold of vulnerability now rose dramatically, especially since a surging tsunami left standing water behind, so sea salt percolated into the soil and rendered land uncultivable for generations. The crowded landscape would also make it much harder for people to escape in large numbers to higher ground.

Thanks to an accident of preservation, we know that tsunamis hit the coast during Yayoi times. About two thousand years ago, rice farmers at Katsukata near Sendai on the northeast coast built extensive rice paddies, which survive in a waterlogged state, so well preserved that both footpaths and human footprints survive. The archaeologists' trenches revealed a thick layer of sand that covers the paddies, deposited by a great tsunami that destroyed rice fields over a large area. Nothing was grown there for four centuries. We have no means of knowing how many people perished in the disaster.

Japan's tsunamis come into sharper focus in later times, when written records provide at least a hint of their frequency.[7] We know that at least 195 tsunamis struck Japan over a period of 1,313 years up to 1997, with such an event averaging every 6.73 years, the highest incidence in the world. A major tsunami followed the Great Hakuho earthquake of November 29, 684 C.E., but no death count survives. The Sanriku earthquake and tsunami of 889 raged ashore in the Sendai region of the northeast and completely destroyed the town of Tagajo, drowning a thousand people. In 1498, six years after Columbus landed in the Bahamas, 30,000 to 40,000 deaths followed a 7.5 earthquake and tsunami. Two centuries later, in 1605, an enormous tsunami caused a water rise of thirty meters affecting much of the Boso Peninsula and Tokyo Bay. At least 5,000 people perished. The litany of death and destruction unfolds inexorably, with a steady increase in the number of casualties. Another tsunami swept ashore on the Sanriku coast in 1896. Thirty-meter waves killed 27,000 people. By this time, the coastal population had mushroomed, with densely packed, growing cities like Tokyo, where an earthquake, tsunami, and fire in 1923 killed over 100,000 people.

How, then, does one protect an increasingly crowded coastline from tsunamis? Back in 1896, after the Sanriku tsunami, which boasted of waves up to forty meters in places, local celebrities encouraged people to move to higher ground. The government did nothing. Inevitably those who had relocated to greater elevations returned gradually to the lowlands where most people lived. In 1933, when another tsunami struck the same shoreline and caused casualties, the government moved forward aggressively with prototypes for tsunami defenses.[8] After further disasters, concrete seawalls came into use, first in Osaka, then elsewhere, at a time when tsunami forecasting was becoming more effective. By the 1960s, construction of coastal defenses was in full swing, most of them dikes up to six meters high, which were considered sufficient to retain minor tsunamis and storm surges from typhoons. However, these proved ineffective against major events. Today, a combination of seawalls, urban defenses, and preparedness on the part of the public appear to be most effective measures, except in the case of really major tsunamis when most bets are off, even when forecasting systems are as close to instantaneous as they can be, like they are today. There are only minutes of warning, even with state-of-the-art communications.

Concrete seawalls, breakwaters, and other protective measures designed to guard against high waves, typhoons, and tsunamis now protect at least 40 percent of Japan's 35,400-kilometer coastline.[9] Most of them lie in areas where the government estimates that there is a more than 90 percent chance that a major earthquake will occur within the next three decades. One of these stretches of coastline lay just where the great tsunami of 2011 struck. Great waves overwhelmed the sea defenses in minutes. Over 20,000 people perished in a catastrophic demonstration of human vulnerability in an overcrowded, urban world.

THE 9.0 UNDERWATER Tohoku earthquake that caused the tsunami shook the eastern Pacific at a depth of thirty-two kilometers on March 11, 2011. The epicenter was about seventy kilometers east of the Oshika Peninsula. The nearest large city was Sendai, about 130 kilometers away. The seabed rose by several meters. Parts of northern Japan moved as much as 2.4 meters closer to the United States. A 400-kilometer stretch of the Japanese coast sank by 0.6 meter, which allowed the resulting tsunami to travel faster and farther inland.

The huge tsunami devastated the coastline of Japan's northern islands. A 670-kilometer stretch of coastline from Erimo in the north to Oarai in the south experienced the full fury of the waves, which overwhelmed defense walls built in anticipation of far lower surge heights. Towns, villages, and ports were destroyed, bridges washed away. Just over an hour after the earthquake, tsunami waves swept ashore and inundated Sendai Airport, sweeping away cars and aircraft and flooding buildings. Motorists on surrounding roads tried vainly to outrun the water but were engulfed in moments as the tsunami flooded the town. Surging water inundated designated tsunami shelters.

It's hard to grasp the devastation caused by an event like this, the powerlessness of humanity against the pitiless sea. Videos shot during the tsunami show the waves racing ashore, piling up, bucking and tumbling houses one against another like pebbles, surging over larger buildings as if breaking against cliffs, then crumbling them. A haze of foam, spray, and moisture rises high into the air like a moving fog bank. Many coastal towns were left little more than piles of collapsed rubble. Survivors wandered in bewilderment and shock through the ruins looking for family and relatives.

Over 46,000 buildings lay in ruins or were totally destroyed, more than double that number seriously damaged. Fishing boats grounded far inland. The damage to infrastructure was immense, including wide-

Figure 10.2 *A tsunami wave crashes ashore at Miyako City in north-eastern Japan, March 11, 2011. Hitoshi Katanoda/Polaris/Newscom.*

spread power failures that led to rolling blackouts, and also to cracks in irrigation dams. Most sobering of all was serious damage to three nuclear power plants, all of which shut down automatically after the earthquake. But tsunami waves swept over seawalls at the Fukushima reactors and flooded the backup diesel generators that provided emergency cooling, which lay rashly close to sea level. A major meltdown ensued; 200,000 people were evacuated; the long-term damage is still being assessed, especially the dangers from radioactive contamination of food, soil, and water supplies. In all, more than 300,000 people were directly affected by the disaster, many of their homes permanently destroyed.

YEARS WILL PASS before the lessons of the Tohoku earthquake and tsunami will be fully digested, for the event raises complex economic, social, and political issues. What changes will be made in building codes? Will the country continue to rely on nuclear power stations? How can

one raise tsunami consciousness among densely packed urban populations? What roles will generational memory and memories of those
killed in the disaster play in the future?

The destruction and loss of life from the Tohoku catastrophe boggles
the mind. But they are dwarfed by devastation caused five years earlier
by the great Indian Ocean earthquake and tsunami of 2004. A submarine earthquake struck early on the morning of December 26, about
eighty kilometers west of Aceh Province in northern Sumatra.[11] Over
200,000 square kilometers of seafloor shifted upward and displaced
billions of tons of seawater. The abrupt rupture triggered a huge tsunami that sent waves hurtling across the ocean to both east and west.
Enormous waves dashed eastward toward Indonesia, Myanmar, and
Malaysia.

Simeulue Island off the Indonesian coast was closest to the epicenter.
A nine-meter tsunami wave struck only eight minutes after the shaking
ended. Directly after the earthquake began, the islanders, wise in the
ways of tsunamis, ran for higher ground. Their villages were flattened,
but only 7 people died out of the 78,000 living close to the island shore.
Cultural traditions extend deep into the past on Simeulue. A severe tsunami had descended on the island in 1907, killing as much as half the
island's population. Oral traditions of the destruction, of the retreating,
then attacking sea, passed into generational memories. The wave is known
locally as a *smong*. Here, oral traditions and the wisdom of ancestors
saved thousands of lives. Hardly anyone else in the path of the tsunami
possessed a similar awareness.

Even as recently as 2004, there were no forecasting devices in place
such as buoys or tidal gauges within 1,600 kilometers of the Aceh epicenter. This is hardly surprising, since most tsunamis occur in the
Pacific, not in the Indian Ocean. Neither seismograph readings from
distant India nor sightings reported by journalists were effective in spreading the word. Scientists at the Pacific Tsunami Warning Center in Hawaii
received reports of the earthquake within fifteen minutes, but issued a report that no reasonable tsunami threat existed. They were wrong, for the
tsunami had already hit Simeulue Island and would shortly arrive on
Aceh's shores.

Five minutes after the earthquake ceased, the sea retreated over three quarters of a kilometer from an Aceh coast crowded with fishing villages. The villagers flocked onto the exposed seabed to pick up helpless fish. Minutes later a deep roar announced the arrival of a 4.9-meter wave, then a 35-meter monster that cleared everything in its path within seconds. Armageddon ensued. Only 400 people out of the 7,500 inhabitants of the town of Lgoknga survived. The same wave toppled villages all along the coast, leaving barren terrain in its wake. Despite friction that reduced its height, the wave climbed as high as fifty meters up cliffs and hillsides. Survivors, who had run for higher ground when the first wave arrived, stood atop hills surrounded with swirling water, corpses, and shattered debris. In places, the second wave surged up to six and a half kilometers inland.

Harrowing scenes unfolded at Lambada, a fishing village near the provincial capital of Banda Aceh. Three booming sounds like gunfire, the retreating sea, then horrendous waves had hundreds of people and their children running frantically ahead of the water. A torrent of fugitives

Figure 10.3 *Damage wrought by the Aceh tsunami at Ulee-Lei beach, Banda Aceh, taken three weeks after the tsunami. Stevens Frederic/ SIPA/Newscom.*

flowed through the streets as the walls of water surged behind them. Tiring children were trampled underfoot. The dirty, debris-filled water pushed people's feet out from under them as they swallowed contaminated water and were gouged by nails, fragments of cars, cycles, and other detritus. Rubble piles occasionally formed islands where bleeding, almost drowned survivors clung for safely. Then the water retreated, sweeping many of them out to sea, never to be seen again. Everywhere the destruction was epochal. About a third of Banda Aceh was leveled, leaving a denuded landscape that extended for over three kilometers. Only 636 people in the affected area survived. Just 40 were women, only 15 children. Over 220,000 people died or went missing in Indonesia. In Banda Aceh alone, 31,000 died.

The tsunami ranged over enormous distances. The Andaman and Nicobar Islands north of Aceh suffered under huge waves that destroyed entire villages and killed about ten thousand Nicobarese, but not the descendants of the most ancient inhabitants. Five small hunting groups, some of the few surviving hunter-gatherers on earth, still flourish on the Nicobars. As soon as the earthquake struck, their elders led the bands into the hills, for their oral traditions had long spoken of shaking ground followed by large waves. The Onge group fled to higher ground when the water level in a nearby creek by their village fell abruptly. None of them perished, whereas forty-five settlers who had taken over Onge land on the flat drowned. Another form of information also helped on Teressa Island, where an employee of India's Port Management Board happened to be an avid watcher of the National Geographic Channel, where he had learned how earthquakes can cause tsunamis. He and a colleague warned five nearby villages. About fifteen hundred people survived as a result.

Another segment of the tsunami wave moved westward and hit the tourist areas at Phuket in Thailand, famous for their beaches. Nearly seventy-five hundred people died, including many European tourists. To the west, the waves hit the east coast of Sri Lanka. Fifteen thousand people perished; eight hundred thousand were rendered homeless. The waves flowed on, doing extensive damage in India and reaching as far as Somalia, where nine-meter waves destroyed several fishing villages and killed nearly three hundred people. Eventually much-reduced waves reached South Africa, even Antarctica.

The 2004 and 2011 disasters provide a graphic look into the future of densely populated coastlines when confronted by a severe tsunami. We've learned how helpless we are, crowded into cities along earthquake-prone coastlines. We've also learned that seawalls are no panacea. Doubtless the Japanese government will continue to erect such structures, for political reasons if nothing else. Unlike typhoons with their storm surges, you cannot provide warning of the arrival of tsunamis days ahead of time, nor will they necessarily strike where they have landed before. Elaborate tsunami warning systems now extend across the Pacific, capable of providing almost instantaneous warnings if an underwater earthquake occurs, but the lead time between warning and sea surge is often minutes rather than hours.

You can build seawalls against gradual sea level rise, as the Dutch have done for centuries, but building them as defense against sea surges and tsunamis is a form of very expensive Russian roulette, with no guarantees of success whatsoever. The Aceh tsunami also taught us that mangroves and floodplain forests reduce the effects of storm surges and tsunamis. In the Aceh case, the greatest mangrove protection came at some distance from the area of maximum impact. A hundred-meter wide stand of mangroves at a density of thirty trees per hundred square meters could reduce flow pressure by 90 percent. In Tamil Nadu, India, villages behind mangrove swamps suffered no damage, while others with cleared swamps had problems. Unfortunately farmers, fishers, and developers cleared half the world's mangrove swamps during the second half of the twentieth century. Now those who unknowingly relied on them for protection are paying the price.

Rising sea levels enhance the dangers posed by underwater earthquakes and tsunamis. What defenses do we have against such cataclysmic events? The vanishing mangroves and tragic experiences of both Aceh and Tohoku tell us once again that the best defense against tsunamis is a human one. Ultimately our only weapon is heightened cultural awareness, something brought home by the persistent and powerful memories of victims who perished in the great tsunamis. Our heavily urbanized societies offer little protection, unlike the ancient strategy of the Nicobar Islanders, whose oral traditions passed from one generation to the next, warning of impending danger, urging flight to higher ground. Today, millions of people living in low-lying coastal areas are like sitting ducks in the face of severe tsunamis—and there is little we can do about it, even with highly effective tsunami warning systems.

Challenging Inundations

Roll on, thou deep and dark blue ocean—roll!
Ten thousand fleets sweep over thee in vain.
Man marks the earth with ruin—his control
Stops with the shore.

Lord Byron, *Childe Harold's Pilgrimage* (1812–1818)

Around 1860, at the heart of the Industrial Revolution, global temperatures began an accelerating climb that continues today and shows no signs of slowing down. With warming came rising sea levels well documented by scientific measurements.

Both warming and climbing oceans are, of course, phenomena that humans have encountered before, but today there's a new element to this familiar experience—rising coastal population. We live in the era of the megacity, of uncontrolled migration from rural hinterlands into ever-expanding urban landscapes. Many of today's cities are chaotic and confrontational. They bristle with seemingly intractable problems of poverty, sanitation, and water shortages, just like the issues faced by our forebears two thousand years ago—but with a difference. Today cities with more than a million inhabitants located close to sea level are now routine. In the next five chapters, I journey repeatedly between past and present, for, in each area we visit, the experience of ancient societies, often our direct ancestors, form a continuity between earlier times and today. We cannot understand the dilemmas of the present without placing them in a deeper historical context. Much of this long-term historical perspective appears in earlier chapters, especially for Bangladesh and the Low Countries, whose stories we round off in coming chapters. In the case of Arctic barrier island communities and small deep-ocean islands, the choices are stark, imminent, and unique—relocation or hugely expensive sea defenses. None of these communities or islands can afford the option chosen by the affluent Low Countries in northern Europe, which are walling themselves off from the attacking sea. These chapters are a chronicle of agonizing options as humanity faces the reality of a fundamental human right—to have enough to eat.

A Right to Subsistence

A COUPLE OF YEARS AGO, I heard retired Bangladeshi major general A. N. M. Muniruzzaman address an environmental conference in Colorado about rising sea levels and the future of his country. He was an articulate speaker, who addressed the audience in a clipped British accent, clearly derived from his days at the Royal Military Academy Sandhurst. He heads the Bangladesh Institute of Peace and Security Studies in Dhaka, an appointment that puts him directly on the horns of the dilemmas facing 168 million of his countrymen, crammed into a muddy, often-inundated delta about the size of Louisiana. The general spared no punches. He couldn't afford to, as Bangladesh's population will reach an estimated 220 million people in a mere forty years. We learned that between 17 and 40 million people living on, or close to, the Bay of Bengal will have to move to other parts of the country in the face of rising sea levels by 2100. The audience gasped. He was talking about tens of millions of environmental refugees not as an abstract problem for the future, but as a sobering reality.

Muniruzzaman believes that these potential environmental refugees are not only a humanitarian problem, but also a serious issue of national security, a subject he knows firsthand. The realities are mind-boggling to the casual observer. There is already unrelenting pressure on agricultural land; millions of displaced people will effectively have nowhere to go in coming decades; the government lacks resources or organizations to handle the potential displacement. Nor is there anywhere for them to go, hemmed in as Bangladesh is by India to the north and west and

Myanmar to the east. Both their neighbors feel deep antipathy to densely populated Bangladesh and possess different religious beliefs. The two countries are understandably nervous about the prospect of uncontrolled mass migrations, which could lead to major epidemics, food and water shortages, and also violence. India is building a four-thousand-kilometer fence to keep out migrants, cattle rustlers, and people seeking work. All of this is quite apart from the nuclear weaponry in the hands of both India and Pakistan, of which Bangladesh was, of course, formerly a part.

Independence from Pakistan came in 1971, after years of ethnic discrimination and political exclusion from a government located nearly 1,600 kilometers away with a far-from-friendly India in between. Cyclone Bhola had come ashore in November 1970, at a time of mounting anger and unrest, on the eve of a national election. Thousands of angry voters, furious at disgracefully lethargic relief efforts, swept the opposition Awami League to victory in the east. Widespread civil disobedience and agitation for independence soon led to the Bangladesh Liberation War, which ended with the formation of an independent Bangladesh in 1971.[1]

Bangladesh is a relatively new country, one with a volatile history and endemic poverty, at great risk from extreme climatic events and rising sea levels. The long history of cyclones and storm surges has resulted in enormous casualties and heart-rending tragedies, but at the same time it has produced a citizenry of tough resilience, whose only long-term weapons against the attacking ocean are low-tech solutions, ingenuity, and human power. Long-term efforts involving both industrialized nations and a wide variety of nongovernmental organizations are under way, projects both to combat environmental problems and to cope with rising population densities. For example, an aggressive family planning program and expansion of nonprofit basic health care have reduced the birth rate and infant mortality dramatically.

Despite continued political turmoil and several military coups, the newly independent government turned its attention to disaster relief, drawing on the lessons of Bhola. The authorities and the Red Crescent (part of the League of Red Cross Societies) cooperated to develop a

Cyclone Preparedness Programme, which came into being in 1972 and is designed to raise public awareness of cyclone risks and to train emergency personnel in coastal regions. The administrators of the program faced a daunting task in a country with a long coastline and a delta landscape that lies but a few meters above today's sea level. Bangladesh is more vulnerable to rising sea levels than any larger nation on earth, with such factors as subsidence and groundwater pollution to combat, as well as a long-term time bomb—massive population growth. The capital, Dhaka, provides sobering numbers. In 1970, 1.4 million inhabitants lived in the city. By 2008, the figure had risen to 14 million. Sober projections speak of a megapolis of 21 million people in 2025. The country as a whole had 44 million inhabitants in 1951. Today a minimum of 168 million people live in one of the most densely populated countries in the world, with 60 percent of them twenty-five years old or younger.

Fortunately, the new government, or perhaps one should say governments in a coup-prone political environment, took the lessons of Cyclone Bhola to heart. One lesson involves the need for efficient early warning systems. As the storm approached, the Indian government had received numerous warnings from ships in the Bay of Bengal. This critical information never reached Dhaka, owing to hostility toward Pakistan. In the end last-minute radio warnings did make about 90 percent of the population aware of the approaching cyclone, but only 1 percent sought refuge in brick structures capable of withstanding the onslaught of wind and water. Enormous casualties inevitably resulted.

Circumstances had changed greatly when another cyclone arrived in 1991.[2] This time, with much more effective warnings, a small army of trained volunteers fanned out over the countryside ahead of the storm to alert even the remotest communities. They evacuated 350,000 people to 508 cyclone shelters before the storm surge submerged over 160 kilometers of the coastline. The actual situation was more complex, as many people either stayed in their homes or moved to embankments or elevated roadways. Much depended on the distance to official cyclone shelters, or the presence of markets, mosques, and schools, which offered refuge in brick-built buildings. Nearly a quarter of all those who did not reach a brick-built structure perished. About 138,000 people

Figure 11.1 *MODIS image of Cyclone Sidr close to the mouth of the Ganges River from NASA's Terra satellite, November 14, 2007. The Category 4 storm is traveling northward over the Bay of Bengal toward the mouth of the Ganges at a speed of 13 kilometers an hour with winds at about 220 kilometers an hour near the storm center. Courtesy: NASA.*

died; 10 million were rendered homeless. But the casualties were far fewer than the half million killed by Bhola.

Relief efforts are now even more effective. When Cyclone Sidr arrived in November 2007, 42,000 volunteers lived in coastal regions, most of them schoolteachers, social workers, imams, and also local government and community leaders.[3] Thanks to more efficient forecasting, improved radio warnings, and the volunteers, Sidr killed only between 10,000 and

12,000 people, far fewer than its predecessors, despite wind speeds in excess of 260 kilometers an hour and a sea surge that reached six meters. This time, more developed roads and better infrastructure allowed the government to evacuate over 3 million people along the coastline. The authorities distributed blankets, tents, and thousands of tons of rice within hours of the disaster, as 700 medical teams moved into position. As one might expect, casualties were highest on the exposed offshore islands, especially among women and small children.

BOTH FLOODS AND sea levels are part of the volatile environmental equation in Bangladesh. Cyclones and storm surges funnel their way up the Bay of Bengal and across the shallow continent shelf that lies immediately offshore. About sixteen tropical cyclones develop in the bay each year, but not all of them affect Bangladesh. When they do arrive, most in May or November immediately before or after the monsoon season, they tend to affect the southeastern portions of the coast. Such events have been part of the delta climate for centuries. In a warmer future, experts predict that there will be significantly more severe storms and sea surges and about 10 percent more monsoon rainfall as well as higher temperatures. With higher sea levels, the effects of the surges will be felt even farther inland than they are today, with an accompanying rise in soil salinity. At the same time, the farmers and fisherfolk of the coast will have to adapt to river floods on a greater scale than in the past.[4]

Most of Bangladesh's rainfall occurs during the monsoon months between June and September, when the country's three great rivers unleash both floodwaters and about a billion tons of sediment on the delta—from a catchment area twelve times the size of the entire country. Such inundations are a fact of life in Bangladesh. The delta farmers are accustomed to seasonal flooding and survive comfortably most years. As much as a quarter of the country vanishes underwater during a normal monsoon season.[5] This figure can rise as high as 70 percent in exceptional years. Very severe floods are another matter and can cause major loss of life, especially today, when population increase is rapid,

cities are growing, and a combination of other forms of economic development and poor maintenance of flood works come into play. Some
authorities have even argued, quite reasonably, that it may be impossible
for Bangladesh to develop a twenty-first-century economy without
proper management of the disaster risks posed by floods—this before
one factors in sea surges and rising sea levels.

Major floods have wreaked significant destruction to crops and the
national economy in at least five years of the past thirty. The floodwaters rise over low riverbanks, break embankments, and flow out over
the featureless landscape. In normal years, many villages are like islands
on low mounds across the plain. Each household constructs an elevated platform inside its house to raise its family above the climbing
water. This works surprisingly well, except when exceptional floods
bring havoc, sweeping everything before them. Hundreds of people routinely die, the damage to infrastructure, crops, and small business often
numbering in the billions of dollars. In earlier times, when population
densities were lower and urban expansion and industrial activity more
muted, much of the flood passed into wetlands surrounding Dhaka
and other cities in a form of natural flood relief. The government has
issued regulations that outlaw infilling for buildings, but developers
blithely ignore them, especially around the capital. Both the poor and
the workers, who then live and labor on these lands, are now at much
greater risk, many of them immigrants from outlying areas who are
unaccustomed to flooding, especially that caused by the sudden breach
in an embankment.

Floods and rising sea levels profoundly affect agriculture. Three quarters of Bangladesh's cultivable land is under rice, most of it traditionally
grown using the monsoon rains, and also the residual moisture in the
ground after the water recedes during the dry season. A big flood erodes
riverbanks and embankments, sweeps away entire villages and their land.
Muddy water can inundate the landscape for weeks, even months, and
devastate traditional *aman* rice varieties that do not like prolonged im-
mersion. During the 1980s, many farmers turned to irrigation agriculture, using a form of dry rice, known as *boro*, which requires a higher
capital investment in dikes and earthworks, but is not dependent on ca-

pricious rains and floods. Higher sea levels near the coast increase soil salinity, but new strains of both crops can be grown under more brackish conditions. While diversifying crops, the government is building up contingency food reserves and food storage facilities to tide people over lean years.

Cyclones, floods, and sea surges remodel the landscape from one year to the next. Even with satellites, computer models, and vastly improved communications, communities living in at-risk areas near the coast, on offshore islands often called *chars*, or near river embankments have to adapt to constantly changing environments. An embankment break or an eroding river can destroy a village and its fields in hours. The inhabitants can do nothing but move to higher ground until the water recedes. They then try to find vacant land nearby to rebuild, usually a difficult task in an increasingly crowded landscape. Sometimes they move onto a wealthier kinsman's property in exchange for some nominal labor. Their old homes are gone forever, but the rivers are so volatile that they can often rebuild on newly formed land that appears without warning as the water falls. But the chances are that they will have to move again. Many landowners become landless, take on mortgages they can ill afford, and live in an endless debt cycle.

However, the basic mechanisms of survival are at least partially codified in law and go back centuries. One can call these people environmental migrants, but almost invariably they move elsewhere temporarily, then return when circumstances permit. They have no desire to leave their ancestral lands, with all the emotional and spiritual ties that underlie the community. Villages on the mainland shift locations again and again in response to flood and sea surge, but they never move far if they can avoid it, even if several members of each household, especially the young men, travel to cities to find work so they can send money home. During severe floods, some families move onto boats. When erosion on an offshore island like Hatia, at the mouth of the great river estuary on the southeast coast, results in the loss of land, the immediate response of those affected is to move to their in-laws' homes, away from the immediate threat. Then they wait for natural accretion of flood deposits to create new land that they can claim.

What will happen in the future, if warmer conditions bring stronger and more frequent cyclones, and Himalayan glaciers far to the north melt faster and swell the great rivers hundreds of kilometers downstream? Most of Bangladesh lies on the fertile alluvial plains of the Ganges and Brahmaputra Rivers. Even with reduced flow, their courses shift constantly, making it hard for the rivers to build up high banks. Would conventional flood works provide a solution? Expensive international schemes proposed by the World Bank envisage building nearly eight thousand kilometers of dikes to control the rivers, at a cost of ten billion dollars, but many local farmers oppose the scheme, as it would result in forced changes to their farming methods. Nor would massive dikes like those in the Netherlands solve the problem, for the subsoil is alluvial sand and mud, which shifts constantly. In any case, Bangladesh lacks the funds to pay for such expensive solutions.

The best solutions appear to lie with the people themselves and with judicious government investments in infrastructure and flood-protected housing. Numerous farmers are building houses on stilts that stand above even the most severe floods. There are smaller-scale solutions as well, among them simple dwellings with cheap jute panel walls that are easily replaced after a flood that are built on half-meter concrete plinths. At the same time, Care and other organizations are encouraging farmers to use long-abandoned farming methods that include floating gardens, well suited to areas that are flooded for long periods of time. Bamboo and dense beds of hyacinth, as well as last year's decomposed vegetation, create floating gardens that allow the growth of vegetables for sale in markets for most of the year, with much higher yields. Salt-resistant rice varieties are another solution, as are fast-growing crops that can be harvested before the monsoon rains arrive. A policy of breaching earthen dikes to encourage silt deposition as the water drains away helps the land rise and counterattacks the effects of rising sea levels.

In the long term, Bangladesh could probably develop ways of living with the unrelenting siege of flood and sea surge. Encroaching water may cause catastrophic damage and kill people, but at least the land is still there when the waters recede, even if erosion rearranges many fields.

But what will happen when rising sea levels take away the land forever? Can the country survive the attacking sea?

SUCH A DIRE prediction is no science fiction. Even the most conservative projections allow for a 10 percent increase in monsoon rainfall by 2050, as the country gets warmer and wetter. Rising temperatures in the order of about 3.5 degrees Celsius will result in greater runoff from Himalayan glaciers. Not only coastal lands but also areas in the middle of the country will be affected. The most drastic effects will result from climbing sea levels, which will encroach dramatically on the flat delta, less than five meters above modern sea level. The past century has seen a rise of about twenty centimeters, which has brought higher soil salinity levels to the near coastal zone. The coastal zone as a whole comprises just over a quarter of Bangladesh's land surface. Nearly a quarter of the country's population lives in this fertile zone, mainly in farming and fishing, and off shrimp farming, which is now a major international industry. The threat of even higher sea levels has come to the fore, as research on global warming has intensified.

Back in the 1980s, a sea level rise of just under 1.5 meters in a century seemed like a reasonable projection. More conservative estimates have extended the timescale to 150 years. If this 150-year forecast became reality, almost 22,000 square kilometers of densely populated coastal territory would be directly affected by seawater, and even be lost to the ocean, a development that would directly impact about 17 million people, about 15 percent of the country's population. Given Bangladesh's exploding population growth, this figure is probably far too conservative.

Long before sea levels actually flood the landscape, another factor is already coming into play—rising soil salinity that is turning much of the southwestern parts of the coastal zone into vast saline swamps where no vegetation grows. Coconut palms and banana groves are dying in the increasingly brackish water. According to soil scientist Golam Mohammad Panaullah, formerly of the Bangladesh Rice Research Institute, 1.5 million hectares suffered from mild salinity in 1973.[6] By 1997, a further

975,000 hectares had been affected. There is no up-to-date survey, but the affected area may now be as much as 3 million hectares. Salinity has risen by about 45 percent in some southern rivers. Saltwater seepage is now penetrating much farther inland, in places in the southwestern portion of the coast as far as 200 kilometers from the Bay of Bengal.

Rising salinity drastically affects soil fertility, to the point that agricultural production has fallen significantly in coastal lands little more than 4.5 meters above sea level.[7] This is especially true when farmers irrigate their land with slightly saline surface water during the low flood months. Every high tide deposits more salt on farmland. Rice production is seriously affected, quite apart from other crops, according to some unofficial estimates down as much as 50 percent in some areas. In the end, much stagnant saline water seeps into the groundwater, making it useless for irrigation or consumption by human or beast. Bangladesh's total rice production may fall by 10 percent and wheat by around 30 percent by 2050.

While the sea encroaches, so river flows are also shrinking, much of it because of much greater use by exploding populations upstream, and also because of damming. The Farakka Barrage across the Ganges, in Indian territory 16.5 kilometers from the Bangladesh frontier, is designed to divert water for Kolkata and to flush out silt from its port on the Hooghly river. Water flow downstream is significantly reduced. Farakka was completed in 1975, but so far the agreements between India and Bangladesh surrounding water release have proved fragile. The reduced water flow is leading to predictable rises in soil salinity in downstream river valleys, where there is now less water to flush out the soil and deposit silt. In the coastal city of Khulna, the main power station depends on freshwater to cool its boilers. A barge goes upstream to collect freshwater for the purpose and now has to go farther and farther upriver for acceptably fresh supplies because of saline intrusion and reduced flow from upstream.

The long-term effects of rising salinity include drastic losses in biodiversity, degradation of fertile agricultural land, scarcities of drinkable water, decreases in freshwater fish populations, and serious threats to long-term food security. The government is well aware of the problem, but has done relatively little to mitigate the situation. Officials point out that farmers could turn to shrimp farming by walling off their rice paddies with high earthen walls to retain saltwater from high tides. Frozen shrimp are a major export industry in Bangladesh, and there is more money in shrimp than in rice. This may be true, but it does not help poverty-stricken rice farmers, who live from harvest to harvest. Most shrimp farming is in the hands of major industrial concerns and wealthy landowners who can afford the high capital expenditure involved in land conversion and maintenance. Many farmers look out over saline puddles that were once fertile rice fields, which they now cross with bamboo bridges and where no vegetation or trees now grow.[8] In a desperate show of resistance to the new order, significant numbers of farmers refuse to sell their land to shrimp companies.

AN ESTIMATED SEVENTEEN to forty million people will be affected if current projections of the effects of sea level rise become reality.[9] Surrounded by saline swamps and useless rice paddies, what will subsistence farmers do? As we've seen, people who lose land to erosion and flood are generally anxious to return to their ancestral homes. If, however, the same land is gradually lost to rising salinity and the ocean, with no prospect of recovery, there remains but one option—to move away completely. The effects are somewhat akin to those experienced by small, low-lying islands like the Maldives or Tuvalu in the Central Pacific, described in chapter 12. In the case of Tuvalu, you are looking at somewhere around ten thousand out-migrants. Bangladesh has tens of millions of potential refugees, under circumstances in which there is effectively nowhere for people to resettle. People in the threatened areas are in quadruple jeopardy, because they have limited income sources, low resilience, and above all, little capacity for adaptation, especially when, as is the case, population growth is rapid and there is great disparity in the distribution of wealth and livelihoods.

Traditionally, people from rural villages have moved to Dhaka and other cities to seek employment. However, cities and towns are already bursting at the seams and there are few job opportunities. Any form of wholesale migration from the countryside is unsustainable, even with massive expenditures on infrastructure and facilities for the people being resettled; there are no funds for such work. Water and sewage problems alone are already monumental. One option might be to decentralize cities and townships in such a way that migration could be organized in a planned way, but this would still not take account of overcrowding in the country generally. Given the slow nature of the environmental changes taking hold in Bangladesh, it's likely that both internal and external migration will be voluntary and, at first, on a relatively small scale. It's when the pressure increases decades from now that political and security problems will mount, especially in a country surrounded by neighbors with whom relations are at times tense. At issue here are each individual's social networks, his or her cultural ability to cope with change, his or her attitudes and position in family and society and gender; all are factors that contribute to the decision to migrate or not.

Cyclones and storm surges are nothing new in human experience. What is new is the extent of damage they can cause and the enormous numbers of people affected by them, often catastrophically. In our warming world, the frequency of such disasters is likely to increase as our vulnerability to the ocean escalates. Bangladesh gives us a haunting snapshot of what lies ahead. At this early stage, when the slow-moving threat is on the horizon and its people are already suffering, we would be wise to abandon any wistful thoughts of business as usual. The impending crisis is not just a Bangladeshi problem but also one that affects us all—for it will. We should remember Article 2.1 of the United Nations covenant, which gives all a right to a "means of subsistence." Rising sea level and their consequences threaten that right in Bangladesh and elsewhere. The defense strategy of last resort, managed resettlement not for thousands but for millions, may move to the front burner.

The Dilemma of Islands

AN ARCTIC OCEAN BARRIER ISLAND, ALASKA. Early summer, 1100 C.E. The kayak drifts slowly in the calm, ice-strewn water close offshore. If the hunter looks to his right, he can see thin plumes of smoke rising from the summer camp on the inconspicuous island. His eyes are never still as he combs the water for a seal coming up to breathe. The ice has gone out early this year, far earlier than it did in his grandfather's day, remembered by tales told on dark winter days. His quarry surfaces by a drifting ice patch. The hunter paddles gently, then drifts as he waits, his harpoon poised for a shot. A ripple in the water, a quick shot: The wounded seal dives as the harpoon float and line mark the spot. Hours later, the kayak returns to shore, towing the carcass of its prey.

By this time, a rising wind is blowing from offshore. Quickly the hunter and his relatives drag the carcass ashore, carrying the kayak to safety. While the women gut, skin, and butcher the seal, the men watch the growing storm and the rising tide. Soon waves are breaking practically underfoot as the ocean cascades ever closer to the hide shelters. Once again, the summer camp is under attack. The hunters quietly prepare their large skin boats, beached on the inshore side of the barrier island, ready to move to higher ground if necessary. Within a few hours, the island is deserted and barely visible above the surf. The people have set up camp elsewhere, but know they'll return to a place where they've hunted for many centuries.

Sea level changes have a long history in the North American Arctic. In about 2500 B.C.E., sea level rise around the Bering Sea slowed, forming beaches along its coastlines. Fishing and sea mammal hunting thrived in the following centuries, as human settlement spread gradually across the far north into what is now eastern Canada. Around the northern shores of Hudson Bay and the southern Canadian Arctic archipelago, some groups settled in areas where game, fish, and sea mammals abounded. As sea levels receded owing to a combination of geological factors, the people moved their camps to stay close to the water's edge. Sea levels declined locally from sixty meters down to four meters above modern sea level.[1] Much of local subsistence came from seal hunting from the ice edge and at winter breathing holes. The hunters would crouch for hours, waiting for the moment when the seal came up to breathe. A quick harpoon thrust, then frantic chiseling with the butt of the weapon to widen the hole in the ice, so they could drag the seal onto the ice before it carried the harpoon, and perhaps the hunter, into the water.

As far as we can discern, coastal populations along the northern shore and islands were extremely sparse, except where caribou herds abounded. After about 700 B.C.E., advances in sea mammal hunting technology, including much more effective harpoons, revolutionized coastal life throughout the Arctic. In the words of archaeologist Owen Mason, life throughout the Bering Strait region was "a welter of small villages with divided and shifting loyalties, multiple origins, and limited spans of occupation."[2] Many communities settled on barrier islands and along low-lying estuaries, where they exploited the seasonal migrations of fish and sea mammals.

Around 1300 C.E., during the warmer centuries of the Medieval Warm Period, groups of highly mobile hunters with kayaks and larger hide boats ranged from the Bering Strait along the Arctic Ocean coast. Their watercraft enabled them to move over wider territories, following seal migration routes through narrow defiles in the ice far from their bases. Ice conditions were less severe during the warmer centuries, so they could also prey on whale migrations that moved eastward in spring, westward in fall, following ice-free leads close to shore. This may have been the time when the ancestors of modern-day Eskimo communities along the Chukchi Sea coast north of the Bering Strait used summer camps on the low barrier islands that protected much of the shore.

Their descendants still hunt on the ice and along the coast, but with a significant difference. Temporary camps have become permanent villages, at a time when global warming melts the ice ever earlier each year.

ACCORDING TO A succinct definition on the web, a barrier island is "a long, relatively narrow island running parallel to the mainland, built up by the action of waves and currents and serving to protect the coast from erosion by surf and tidal surges."[3] Barrier islands often develop in the mouths of flooded river valleys as the sea levels rise, or at river mouths where sediment accumulates and forms a delta. Local geology, sea level changes, vegetation, and wave activity are but a few of the factors that can affect such islands. Over the past five thousand years, rising sea levels have created a plethora of barrier islands, especially in the Arctic and North Atlantic. They are rarer in the Southern Hemisphere, where sea levels have tended to be more stable. Over two thousand barrier islands are known, many of them identified from high quality satellite imagery. They most typically form along geologically stable coasts with shallow estuaries, like those along much of the eastern United States. About three quarters of the world's barrier islands are in the Northern Hemisphere, most of them in the high-latitude Arctic, where sea levels are rising faster than anywhere else.

Arctic barrier islands are more vulnerable to climate change than any other such formations in the world. For thousands of years, permafrost and sea ice have protected them from wave damage during storms. Now sea levels are climbing aggressively by geological standards. Islands in the far north are said to be eroding three or four times faster than those farther south in the United States. If the erosion accelerates, many Arctic islands will vanish and the Eskimo communities living on them will face an uncertain future.

Today, most villages along the Chukchi Sea coast are clusters of heavily insulated government-issue homes that protect their occupants against a climate where ice is at their doorsteps for nine months a year—or was until recently. Today, ice-free conditions along the shore last for as long as four months, sometimes more, along Alaska's North

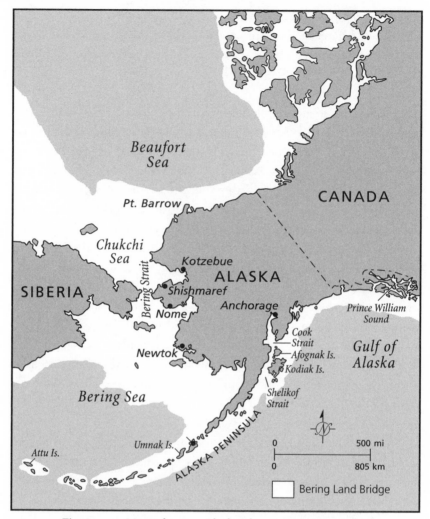

Figure 12.1 *Maps showing Alaskan locations in chapter 12.*

Slope, even longer farther south. Now fall and winter storms bring waves
that break against the fragile barrier islands, smashing the melting permafrost that once helped make the beaches natural seawalls. A combination of rising sea levels and greatly accelerating erosion are playing havoc
with the communities that have used Arctic barrier islands for a very long
time. The US government estimates that at least twelve Native American

villages are facing possible destruction. Another twenty-two coastal communities will require some form of immediate protection from climbing sea levels and its consequences. The people of these villages, who are still subsistence hunters, once moved easily in the face of changing climatic conditions and the seasons. Living permanently in places that still make sense in terms of hunting strategy, but are not necessarily the best places, has made these isolated communities highly vulnerable to rising sea levels in ways that were unimaginable in ancient times.

Shishmaref, on the shores of the Chukchi Sea, lies on a six-kilometer-long island, a settlement occupied by 580 subsistence hunters. People have visited here for at least four thousand years, probably much longer. Until recently, the Eskimo used the island as a winter camp, fanning out to other island encampments in the summer.[4] Toward the end of the nineteenth century, Shishmaref became a harbor for shipping gold-mining supplies inland and permanent occupation began. Now the community is under serious threat from rising sea levels, with only limited options for long-term survival.

Figure 12.2 *A house falls into the sea at Shishmaref, Alaska, September 27, 2006. AFP Photo/Gabriel Bouys Files/Newscom.*

One option has been in play for a while—constructing sea defenses. Since the 1950s, the community has tried a variety of measures, including oil drums and sandbags, even household refuse, to protect the settlement against storms. In 1984, Shishmaref built a 520-meter seawall of wire baskets filled with stones. Storms promptly removed sand behind it, but the wall did slow coastal retreat. Next came a barrier of cement blocks, linked by cables to form a mat and placed against the face of the coastal dune, designed to bend when sea ice pushed against it. The mat failed in short order. Even if successful, the walled area actually enhances the problem, for the sea is cutting back on both sides of it and turning the village into a headland and a juicy target for storm waves. By late 2006, some thirty-four million dollars had gone into seawalls to protect the community, a staggering expense to protect less than six hundred people and more costly than all the structures in the village.

Then there's a second option: Remain on the island and shift threatened homes to new locations. Eighteen houses were moved back from the shoreline after a 1997 storm. However, wherever one places houses on the island, they are still threatened by ongoing erosion, so the only path for long-term survival would be a community protected by prohibitively expensive high seawalls. Such an adaptation scenario would also require the construction of storm shelters and the preparation of evacuation plans that could be implemented in very harsh weather and at short notice.

A third option: Move to neighboring villages or to larger communities such as Anchorage, Kotzebue, or Nome. Joining nearby subsistence-based villages would place local food resources under threat, apart from the complexities of traditional rivalries, some of which go back generations. Migrating to cities or towns would mean the immediate loss of a subsistence lifeway that dates back many centuries and is the last surviving element of a once-thriving culture. Both Kotzebue and Nome, with populations in the range of three thousand to four thousand, are prepared to accept the villagers, but how would the hunters and fisherfolk support themselves in their market-driven economies? To add to the complications, they would come from a totally dry community to towns that have serious problems with endemic alcoholism.

There remains a fourth option: Move Shishmaref and other threatened barrier island communities to higher ground. The idea was first mooted as early as 1973, but it was not until 2002 that the villagers voted to move to a new site over a period of years. Shishmaref's new site on the mainland is known as West Nantuq, across a lagoon. An Army Corps of Engineers study in 2004 estimated the cost of relocation at $180 million, including moving 137 homes across the frozen lagoon in winter and bringing in some additional prefabricated houses by barge. The federal government is expected to pay for everything, including the complete infrastructure for the village and a new harbor. Shishmaref may have ten to fifteen years before it vanishes. Meanwhile, relocation plans move along glacially slowly.

Other coastal Alaskan villages also wrestle with rising sea levels, among them Newtok on the Ninglick River considerably farther south, a place close to the Bering Sea that fisherfolk and hunters have visited for at least two thousand years. Some three hundred Yupik Eskimos live in the village, which is rapidly being washed away by erosion caused by the ever-widening river. The villain is rising temperatures, which have reduced ice cover, brought more frequent storm surges, and thawed the permafrost, which formed a buffer against the Bering and formed a solid foundation for Newtok. Now the underpinning is turning to squishy mud; buildings including the old school and community church have buckled and are sinking. Wooden boardwalks connecting the buildings literally float on the muddy ground. The river gobbles up as much as 27.5 meters of dry land each year. Newtok is below sea level and sinking, an island caught between the Ninglick and a nearby slough. The village will have vanished within a decade or so.

Newtok has but one future—to move to a site on Nelson Island, 14.5 kilometers away, which the villagers acquired in 2003 through a land swap with the U.S. Fish and Wildlife Service. They named the new settlement Mertarvik, "getting water from the puddle." The community then obtained funds for building a barge landing for offloading the materials needed to construct the infrastructure for the new settlement. Infrastructure is being installed slowly; house construction started in 2011. The cost of the move is unknown, but government estimates of two million dollars a household seem absurdly high.

There are no normal provisions in federal or state budgeting for funds for a community facing not the devastation of a fast-moving climatic event like Hurricane Katrina, where emergency funding was available in short order, but rather a slow-moving disaster that will eradicate a village of isolated subsistence farmers after decades of slow death. None of this will be cheap and none of it easy for the local people, who are severing ties with village sites where their ancestors lived for many centuries.

At least Newtok, Shishmaref, and other Alaskan villages can move to nearby higher ground—if someone will pay for the relocation. But what happens when your home is completely surrounded by water and lies only a few meters above sea level, as is the case in the island nations of the Pacific and Indian Oceans?

LIVING ON A Polynesian atoll, you can never tune out the sound of ocean breakers. You are at most some four or five meters above sea level. When severe storms descend, you are almost certain to get wet. To live sustainably on such isolated and tiny spots of land in the past required not only water but also a set of adaptations. As a result, every inhabited low Polynesian atoll is a humanly modified environment, complete with seawalls and pits for growing taro, a staple root crop in the South Pacific, the soils often formed of the refuse from human occupation. A few plants have always provided food and essential raw materials— coconuts, with their fluid, meat, and leaves for weaving, pandanus for both food and leaves. Both do well in the salty environments of tropical beaches. Breadfruit and taro can be grown, both requiring water, and the latter grown in artificial pits mulched with leaves and other organic detritus. Inevitably, survival depended on social contacts and trade with other islands, maintained by long-distance voyaging in outrigger and double-hulled canoes.

Both Tuvalu and Kiribati, described below, are Polynesian outliers, settled not from the west like other Micronesian islands, but from the south, from the heart of Polynesia. Intensive studies of radiocarbon dates from Polynesian islands place the first voyages eastward from Fiji and Samoa at around the eleventh century C.E. Both Tuvalu and Kiribati may have received their first inhabitants at about that time: We don't know.[5]

Once settled on a small island, however remote, you were never stuck in one place. There was a great deal of movement in Tuvalu and Kiribati's world, for no island in this part of Polynesia was completely isolated from others over the horizon. That was the strength of Polynesian society and its greatest weapon against an attacking sea—the ability to move elsewhere on short notice. Today, frontiers set over the past century by colonial powers create artificial barriers at a time of fast-rising populations. The ancient, more flexible world is no more. Many Pacific islands face an uncertain future as independent nations in an intensely competitive, global world.

Tuvalu comprises four reef islands and five atolls located in the heart of the Pacific midway between Australia and Hawaii.[6] At twenty-six square kilometers, Tuvalu is the fourth-smallest nation on earth. Only Monaco, the Republic of Nauru, and Vatican City occupy less area. There are 10,500 inhabitants on eight of the nine islands. The highest point lies only 4.6 meters above sea level.

Some canoe loads of people had settled on Tuvalu by at least a thousand years ago, probably from Fiji or Samoa. Sporadic European contact began when the Spanish navigator Álvaro de Mendaña de Neira sailed through the archipelago in 1568 while on a search for the mythical Terra Australis, the great southern continent, but he was unable to land. Foreign visitors were rare until the nineteenth century, except for the occasional whaler, but landing was always difficult. Nevertheless, slave traders, "blackbirders," removed nearly four hundred men from the islands to work in the notorious guano mines on the Chincha Islands off the Peruvian coast between 1862 and 1865. By 1865, Christian missionaries and foreign traders were active on the islands, which became part of the British colony of Gilbert and Ellice Islands from 1916 to 1974. Tuvalu

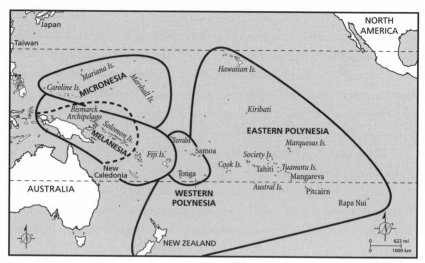

Figure 12.3 *The Pacific Islands.*

became an independent nation within the British Commonwealth in
1978, but it is a nation at serious environmental risk, lying as it does
only a few meters above an inexorably rising Pacific.

The Tuvalu government is profoundly concerned about global warming and rising sea levels, which threaten to submerge the entire country.
Even without sea level rises, the islands are extremely vulnerable to destructive wave action. For instance, the construction of a World War II
airfield on the Islands of Funafuti involved the building of piers, infilling
beach areas, and excavating deepwater access channels. These humanly
made alterations in the 1940s changed local wave patterns, so much so
that much less sand now accumulates to form and replenish beaches.
Hungry waves now devour and erode the shore even faster than before.

Apart from rising sea levels, there are also concerns about so-called
king tide events, exceptionally high tides that occur at the end of the
southern summer and raise sea levels above normal high tide limits.
Even without factoring in rising sea levels, such tides already flood low-
lying areas, including the airport. A belt of narrow storm dunes lies on
the ocean side of the islands, the highest ground on Tuvalu. But occa-

sional cyclones topple these formations, again increasing the islanders' vulnerability to rising seas—and more frequent severe weather events are forecast for the future. High tides and cyclones cause enough disruption before one factors in an estimated sea level rise of twenty to forty centimeters over the next century, which could render Tuvalu uninhabitable.

With the population having more than doubled since 1980 and a history of poor coastal management, Tuvalu's future as a self-sustaining country is, at best, uncertain. Small islands like Tuvalu have few options. There is nowhere to retreat to, nor is it even marginally economical to build seawalls or to reclaim land in the face of the encroaching ocean. There remains relocation, perhaps moving to another island or to Australia, New Zealand, or elsewhere where there is space to absorb thousands of islanders. Just as in other places, there are deep ancestral ties to the islands, which means that at present very few Tuvalans leave their homeland permanently, perhaps no more than seven people out of every thousand. Seasonal employment in agriculture in New Zealand is accessible to up to five thousand workers from Tuvalu and other Pacific islands, with talk of expanding the scheme to Australia, but this does not offer a permanent solution for a vanishing archipelago.

Some commentators have called for the relocation of the entire population to Australia, New Zealand, or to Kioa Island near Vanua Levu, a major island in the Fijian archipelago. Kioa has been a freehold for settlers from Tuvalu since 1947. However, a mass migration to the island bristles with economic and political difficulties, and is not on the immediate horizon. Former Tuvalu prime minister Maatia Toafa has pointed out that the government does not consider the threat sufficient to evacuate everybody. Of more immediate concern is a serious shortfall in freshwater supplies, especially during droughts caused by La Niña. A long dry spell in 2011 led to water rationing, households on the capital atoll of Funafuti being limited to two buckets a day. Desalinization plants are likely to be the dominant water source in future centuries—if the islands are still inhabited. Fortunately, unlike some other Pacific islands, Tuvalu has a guarantee from New Zealand that land and space will be found for everyone on the islands if such a dire situation as permanent abandonment arises.

THIRTY-TWO ATOLLS AND a single raised coral island comprise the tiny nation of Kiribati in the central Pacific, none of them more than two to three meters above sea level.[7] Nearly 113,000 people live on twenty-one of these islands, scattered over three and a half million square kilometers, most of them on the principal island, Tarawa, where there is severe overcrowding. One of the poorest nations on earth, Kiribati is on the frontlines of global climate change, for sea level rise may submerge the islands within a few generations. Here, the coast *is* Kiribati. In an era of more frequent storms and sea surges, also higher waves, the islands are under constant attack. Official tidal records chronicle a sea level rise of about three millimeters a year since 1992, but such readings can be problematic as periodic El Niño events raise sea levels, while La Niñas bring smaller rises or even slight falls.

Rising seas are but part of the problem. The extreme weather events that accompany them are far more damaging, if a projected fifty-centimeter rise—perhaps a conservative estimate—transpires. Coastal erosion is already pervasive. High waves, some 3.5 meters high, sweep over seawalls, topple coconut and papaya trees, and flood houses and gardens. A World Bank report projects that between a quarter and a half of the southern portions of Tarawa, where over half the population lives, will be inundated by 2050. As much as 80 percent of the northern areas could vanish under rising sea levels and storm surges in the same time frame. The measured statistics provide sobering data, but they also mask a much more volatile situation. Events such as exceptionally high king tides are sweeping farmland out to sea and devastating villages, as well as contaminating freshwater wells. People who have lost their houses to the ocean are moving to slightly higher ground, but all too often there is nowhere to go. Saltwater intrusion is already affecting taro crops, which are sensitive to changes in groundwater, as are coconut trees, which provide copra, dried coconut meat, a major component in Kiribati's economy. As the land area diminishes in the face of the Pacific and seawater contaminates wells, the islanders face serious water shortages, even after building a network of boreholes that are fast drawing down an underground freshwater lense under Tarawa, formed by percolating rainwater.

Kiribatans have a profound attachment to the land of their ancestors, but they face a future of limited options like their distant neighbors on Tuvalu. The islanders have adapted about as far as they can, given a population density on Tarawa of about three times that of Tokyo, around fifteen thousand people per square kilometer. There are some seawalls, but to construct the kinds of massive defenses that would be needed to curb storm surges is unaffordable. There is but one long-term solution: out-migration, organized not haphazardly, but in a controlled manner. Talk of constructing artificial islands like oil rigs at an estimated cost of about two billion dollars each went nowhere. The government is now talking of purchasing about 2,400 hectares of land on the island of Vitu Levu in Fiji to relocate environmental refugees.

Rising sea levels are part of a much more complex problem, something so imperceptible that many Kiribatans do not consider it a threat. It is, after all, hard to believe that much of Kiribati, or Tuvalu for that matter, will be underwater within less than a century. Rapidly growing island populations, contaminated water supplies, and the loss of agricultural land to exceptionally high tides and sea surges are immediate concerns. So are flooded houses and villages, and receding coastlines. Food resources are under stress. A combination of all these factors limits the islanders' options severely, to the point that in the final analysis possible solutions all revolve around out-migration, however unwilling Kiribatans are to face it. Where will refugees go? At this point, no one knows.

The Australian government is supporting the training of small numbers of young Kiribatan people in employable skills like nursing, on the assumption that they will remain in Australia instead of returning to their overpopulated, threatened homeland. The president of Kiribati supports this program, which he calls "migrating with dignity." Such limited schemes provide economic incentives for people to move away gradually so they have time to adapt and build Kiribatan communities elsewhere. However, the migration issue is far larger than a few training programs, and is compounded by the refusal of many islanders to believe that there is a sea level problem. Many of them, devout Christians, believe that God will raise the islands a little higher, then higher, so that they, and their descendants, will live in peace.

KIRIBATI AND TUVALU are but two of many Pacific islands that face uncertain futures in the face of the ocean. Such islands have formed an intergovernmental organization of low-lying coastal and small island nations, the Alliance of Small Island States.[8] This active lobbying group argues that serious attention should be paid to mitigating the effects of greenhouse gas emissions and urges support for its members' attempts to mitigate the effects of rising sea levels and other facets of global warming. Fourteen Pacific island states are members, as are the Comoros Islands, Mauritius, the Seychelles, and the Maldive Islands in the Indian Ocean.

The Maldives, a nation of about 1,190 islands covering an area of ninety thousand square kilometers, are a tourist paradise of dazzlingly clear turquoise waters, shimmering white beaches, and pristine coral reefs. The islands lie no more than 2.4 meters above sea level. Some are coral atolls, others islands covered with lush tropical vegetation and reef-ringed lagoons. Three hundred thousand people live here, on some of the lowest inhabited land on earth.

For many centuries, the Maldives were a crossroads astride ancient Indian Ocean trade routes that linked them with India and Sri Lanka. Maldive cowrie shells became a major form of barter currency throughout much of Asia and as far afield as the East African coast for many centuries.[9] Little is known of the earliest inhabitants, who probably arrived from India or Sri Lanka and appear to have lived in transitory settlements. Buddhism spread to the Maldives during the third century B.C.E., probably during the reign of the Mauryan emperor Ashoka the Great, when the religion diffused far beyond the boundaries of his empire centered on the Ganges River. Fourteen centuries later, Buddhism gave way to Islam, which arrived from India's Malabar Coast, a hub of the Indian Ocean trade. Islamic interest in the islands revolved around the cowrie shell trade. Portuguese merchants established a small trading post supervised from Goa on the Indian mainland in 1558, but they were soon driven out. Both the Dutch, who had replaced the Portuguese as the dominant power on Sri Lanka, and the British, who expelled them in 1796, established hegemony over the Maldives, but did nothing to involve themselves in local affairs. Eventually the islands became a British Protectorate from 1887 until 1953, when the Maldives achieved independence and became a republic rather than a sultanate. Since then, the islands' history has occasionally been turbulent, its economy boosted by a burgeoning tourist industry that capitalizes on the great natural beauty of the islands. But that beauty is under siege from climbing sea levels.

At the Kyoto climate change conference in 1997, former Maldives president Maumoon Abdul Gayoom said, "What you will do or not do here will greatly influence the fate of my people. It can also change the course of world history."[10] One cannot imagine the inundation of the Maldives Islands changing world history, but no one can deny that the nation's 2.4-meter-high domains are extremely vulnerable, not only to rising sea levels, but also to exceptionally high tides.[11] In 1987, such a tide flooded Malé, the capital. Nineteen years later, and at huge expense, then-president Gayoom increased the height of Hulhumalé Island as a place of refuge by using a giant dredge to suck up sand from the ocean floor into a shallow lagoon. He built a hospital, apartments, and government buildings. Several thousand people live there: Gayoom wanted fifty thousand to make it home. Rare events, like the great Indian Ocean tsunami of 2004, also cause havoc on low-lying Maldivian atolls. Eighty-two people died and twelve thousand were displaced by the surging waves, which also did millions of dollars of damage to luxury resorts. In 2008, then-president Mohamed Nasheed announced a plan to purchase land in Sri Lanka or India to resettle the Maldive Islanders, but nothing has yet come of this idea.

No one knows what will happen when sea levels rise to the point when the Maldives lose significant land. The three hundred thousand people living on the islands—and the population is increasing—are already finding themselves competing for ever-shrinking space, while,

Figure 12.4 *Malé, capital of the Maldive Islands. J. W. Alker. © Image-broker/Alamy.*

inevitably, the tourist economy will slowly implode in the face of the ocean.[12] Maldivian fisheries, a staple of the economy and a primary food source for the islanders, are already in decline, thanks to pollution and coral destruction. The options for the future are at best limited, for expensive sea defenses or even floating islands are far beyond the resources of the government, even with substantial foreign assistance. Will there be violence in a country short of land where authoritarian rule has a long history, despite democratic experiments? Or will the entire population have to leave the sinking islands and settle elsewhere? There is even talk of the country suing the United States for not doing enough to combat global warming and rising sea levels. In the real world, however, the islanders will probably continue to occupy their homes until the bitter end, for to move away is always heart wrenching and to be avoided at all costs.

Herein lies the dilemma of small islands: Their options are far more limited than people on low-lying coastal plains, who at least have more space. There are no utopian solutions for people living in the midst of open ocean where swells can build over thousands of kilometers. It is not as if they are living on lakes, like Lake Titicaca, where the Uros people used to live in forty fishing villages fashioned from totora reeds that float in the middle of the lake. Most of the Uros now dwell on the mainland, free of the constant labor of keeping their homes afloat.

Small island governments are vociferous about their plight, often blaming industrialized countries for global warming and rising sea levels, and pushing for financial assistance to combat their impending inundation. Such lobbying seems to have fallen on largely deaf ears, for the problems of remote island nations are hardly a high priority in the international realm, especially when the solutions involve long-term thinking that is often unimaginable to politicians obsessed with election cycles. However, as sea levels slowly rise, extreme weather events with their storm surges become more frequent, and ever-lower islands face death sentences from unpredictable tsunamis, the prospect of large-scale out-migration must be considered just as in Bangladesh.

The crisis of barrier islands and remote atolls unfolds at a time when the world lacks any international policies for coping with climatic refugees from small island nations, let alone people in river deltas and on mainland shores. Their plight, still coming into urgent focus, is a global challenge that cries out for measured, international policies to confront the problem head-on before thousands of disoriented, fearful, and hungry migrants arrive on foreign shores where they are unwelcome. For the first time in history, we face a challenge of forced involuntary migration for millions of people, triggered not by ruthless kings, conquering armies, or prosecuting fanatics, but by the natural forces of our capricious world.

"The Crookedest River
in the World"

MARK TWAIN MEMORABLY OBSERVED that "the Mississippi . . . is not a commonplace river, but on the contrary is in all ways remarkable." He also called it "the crookedest river in the world, since in one part of its journey it uses up one thousand three hundred miles to cover the same distance that the crow would fly in six hundred and seventy-five."[1] This is a stupendous river by any standards, 3,779 kilometers long, with a huge triangular drainage area that covers about 40 percent of the United States, the third-largest river drainage in the world, exceeded only by the Amazon and the Congo. The river rises in Lake Itasca, Minnesota, and then flows through the heart of the Midwest. The Missouri River with its vast silt load, the Great Muddy, which drains the Great Plains, joins the Mississippi at St. Louis, the Ohio River at Cairo, Illinois. Below Cairo, the river flows through a wide, low valley, once a bay of the Gulf of Mexico, now filled with sediment. Today 966 kilometers downstream the Mississippi joins the Gulf. The river channel meanders over the low-lying plain, often contained between natural levees formed by flood sediments.

Through a natural process known as avulsion, literally delta switching, the river has meandered back and forth across the flat landscape of the Lower Mississippi valley ever since sea levels climbed in the Gulf of Mexico after the Ice Age. The river's gradient shallows with the ocean's rise, the flow slows, and the silt load sinks to form lobes of a huge delta in a regime that changed little for thousands of years—until humans started controlling the Mississippi. Sediment builds up; a channel

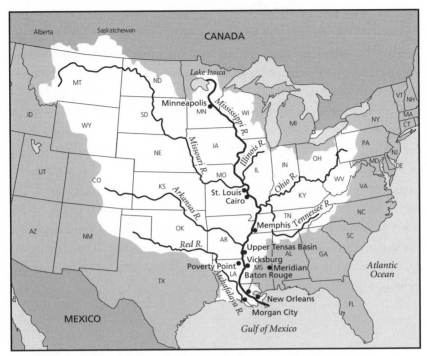

Figure 13.1 *Map showing locations in chapter 13.*

becomes clogged; the river shifts course to a steeper route downstream. Meanwhile the abandoned channel receives less water and becomes a bayou. A major channel shift triggered by an unusually severe spring flood takes place about every thousand years. The last one would have inundated maize fields on the floodplain, but the ancient farmers, hunters, and fisherfolk on the flatlands and among the bayous would have adapted to the shift without trouble. Today's river is long overdue for a dramatic channel shift, most likely down the Atchafalaya Basin or through Lake Pontchartrain near New Orleans. This time the human and economic stakes are very high indeed, with almost unthinkable consequences for cities like New Orleans and Baton Rouge. Only massive humanly constructed flood control works stand between millions of people and disaster from upstream.

Most of the Lower Mississippi's water comes from the Ohio River and from downstream tributaries like the Arkansas and Red Rivers, only 15 percent from its own upper reaches. The Missouri contributes considerably less water, but massive quantities of silt, both of which flow down the Lower Mississippi. All these sources make for a complex flood regimen, especially when all major tributaries overflow at the same time. Such events resonate in historical memory. The Great Mississippi Flood of 1927 produced overflows so severe that the river reached a width of ninety-seven kilometers. On April 15, 385 millimeters of rain fell on New Orleans, covering parts of the city with more than two meters of water. The Flood Control Act of 1928 authorized the Army Corps of Engineers to construct the longest levee system in the world. With the new levees came at least some protection from raging floods, but also a new ever-present threat of highly destructive breaks in the defense walls caused by floods and hurricane-induced sea surges. For all the additional protection, many communities in the shadow of levees were potentially even more vulnerable than before.

THE ROUTINE NEVER changed—enormous flocks of migrating waterfowl flying northward in spring, south in fall, along what is now known as the Mississippi flyway. Thousands of birds would pause to feed and rest at shallow oxbow lakes near the great river. Each spring and fall, the hunters would wait in the reeds at dawn with traps and spears. They would use canoes to drive the birds into narrow defiles in the reeds, where they could be netted, or swim among them with duck decoys on their heads, then grab their unsuspecting prey by the feet from underwater. Back ashore, they preserve the birds by drying and soaking them in oil for later consumption. Storage was critical. Everyone living along the river had experienced food shortages caused by floods that could inundate wide tracts of the floodplain until as late as July.

Between about 4500 and 4000 B.C.E., the Mississippi and its tributaries slowed as sea levels stabilized. Silt from the now more sluggish river accumulated. Backwater swamps and oxbows formed, which proved to be a paradise for hunters camped along their floodplains. Apart from waterfowl, they thrived off fish and mollusks, also plant foods, which abounded, especially the nut harvests of fall. So plentiful were food supplies that many groups stayed in the same places for most, if not all, the year.

By 2000 B.C.E., the Lower Mississippi had become a complex political and social world. Most people now lived in small base camps, but sometimes clustered around somewhat larger centers, connected to other groups by intricate ties of kin and volatile rivalries. The largest of these centers comes as somewhat of a surprise in a world of hamlets and temporary camps. The great horseshoe-shaped earthworks and mounds of Poverty Point lie on the Macon Ridge in the Mississippi floodplain, near the confluences of six rivers and twenty-five kilometers from the great river itself. Six concentric semicircular earthen ridges divided into segments lie about forty meters apart. Apparently houses lay atop the earthworks, which were about twenty-five meters wide and three meters high, elevated above the surrounding low-lying terrain. To the west, an earthen mound stands more nearly twenty meters high and two hundred meters long. Over five thousand cubic meters of basket-hefted soil went into the Poverty Point earthworks.[2]

Figure 13.2 *The earthworks at Poverty Point, Louisiana. © Martin Pate.*

The concentric earthworks came into being in about 1650 B.C.E., surrounded by a network of lesser centers. Long-distance trade routes carrying exotic materials converged here, not only from upstream along the Mississippi, but also from the Arkansas, Red, Ohio, and Tennessee Rivers as well. Together, they formed the nexus of a vast exchange network that handled exotic rocks and minerals such as galena from more than ten sources in the Midwest and Southeast, some of them as far as a thousand kilometers away.

Poverty Point is a huge enigma. How many people lived there? Was this a center where hundreds of visitors gathered for major ceremonies, perhaps on occasions such as the solstices? A person standing on the largest Poverty Point mound can sight the vernal and autumnal equinoxes directly across the center of the earthworks to the east. This is the point where the sun rises on the first days of spring and fall, but whether this was of ritual importance remains a mystery. What is certain, however, is that Poverty Point was at the mercy of river floods and the vagaries of the Mississippi delta downstream.

The great center lies near natural escarpments, close to floodplain swamps, oxbow lakes, and upland hunting grounds. Like societies up and down the major rivers nearby, Poverty Point people were hunters and plant gatherers. They also cultivated a number of native species, such as sunflowers, bottle gourds, and squashes. The floodplain landscapes and their environs produced more than enough food for considerable numbers of people to live permanently in places as large as Poverty Point, but we still know little about them, or of the leaders behind the centers, who must have organized the communal labor needed to erect earthworks and mounds. Poverty Point was a prophetic forerunner of much more elaborate chiefdoms that flourished along the Mississippi and its tributaries a thousand years later. But, after centuries of gradual population growth and extensive long-distance trade, Poverty Point society gradually imploded. The exchange system collapsed after 1000 B.C.E. Three hundred years later Poverty Point was deserted. Everyone had moved away. Trade slowed, settlements were smaller, and any semblance of complex society vanished. One major factor may have been climate change.

Tracking climate change in the Lower Mississippi valley is an exercise in complexity, for riverbank breaks, shifts in large meander belts, and other geological processes resulted in major changes in human settlement, especially between about 1000 and 450 B.C.E. During these centuries, rainfall was higher and temperatures were cooler over wide areas of the world, higher rainfall notably causing increased flooding in the Netherlands, with abandonment of many low-lying areas. We know that cooler and wetter climate pertained in some parts of Minnesota and that there were larger than normal floods along Mississippi tributaries in southwestern Wisconsin. Out in the Gulf of Mexico, the Orca Basin traps sediment from the great river, whose varying size reflects the intensity of ancient floods. At least two flood cycles of unprecedented volume occurred between about 1000 and 550 B.C.E., which introduced large quantities of freshwater into the Gulf. Each of these major cycles lasted for as long as fifty years and must have had major effects on the hydrologic system of the Mississippi River basin. At the same time, between about 900 and 550 B.C.E., many more hurricanes and large storms came ashore along the Gulf Coast. The combined effects of lower temperatures, higher rainfall, and changed atmospheric circulation produced much more flooding along the Mississippi, with resulting serious disruption of daily life among the societies along its banks, perhaps the explanation for a sharp decline in the number of archaeological sites and possibly in human populations in the alluvial portions of the Mississippi River basin as Poverty Point was abandoned.

During the time when Poverty Point prospered, before 1000 B.C.E., the river enjoyed a period of relative geological stability.[3] Around then, the Mississippi channel shifted north of Vicksburg, Mississippi, upstream of Poverty Point, causing westward migrations in river channels downstream, including Joe's Bayou near Poverty Point, which became a major outlet for the river, just as major flooding affected the region. The floods rendered much of the alluvial plain uninhabitable for long periods of time, sweeping away ponds and sloughs where the people had harvested thousands of fish as floods receded. Fish densities may have plummeted in the now faster-moving water, which may have forced the groups who had relied on the fish harvests to rely more heavily on hunting and plant foods in the nearby uplands. No longer could people live in permanent settlements; the floods would have disrupted long-distance trade by canoe.

Mobility versus sedentary living: Poverty Point's demise may be a classic example of increased vulnerability to rising water on the part of more permanent settlements. The change was not instantaneous; people and individuals respond to climatic events in complex ways. As more elaborate ancient societies came into being in later times, it's noticeable that many villages lay on slightly higher, better-drained ground above the floodplain, even if they were near it. Over many centuries, people living along the river and in the bayous and swamps of the delta adapted effortlessly to a regime of flood and sea surge, an equation that changed dramatically with the arrival of European settlers. The cumulative effects of human efforts to control the river and its capricious flooding have now brought rising sea levels to the forefront in the complex minuet of ocean, river flood, and silt accumulation.

THE DANCE FLOOR is the shallow delta that forms the mouth of the river. As the Mississippi approaches the Gulf Coast, the current slows and deposits fine alluvium. Over the past five thousand years, the southern coast of Louisiana has advanced between twenty-four and eighty kilometers, forming twelve thousand square kilometers of coastal wetlands and extensive tracts of salt marsh. The latest cycle of delta development began as sea levels rose at the end of the Ice Age some fifteen thousand years ago. At the time, the mouth of the river was farther out in the Gulf, the current shoreline beginning to take shape about five thousand to six thousand years ago, when sea levels stabilized somewhat. Since then, the river has changed course repeatedly, as shorter and steeper routes to the Gulf have emerged upstream. As the river shifted, so the old delta lobe into the Gulf would lose its supply of alluvium, then compact and subside, retreating as the advancing ocean formed bayous, sounds, and lakes.

The Mississippi delta has always been a landscape of shifting currents, tides, and mudflats. This is a waterlogged, now-vanishing world, where you canoe through forests and occasional open clearings, where there is almost no solid ground. A confusing mosaic of forested basins and streams form the fretwork of the delta, which absorbs the flood and holds the water until it escapes to the Gulf. When floods spread water out over the surrounding countryside, the water slows and deposits silt, the finest dropping at the edges, closer to the river, where natural levees form.

Such levees were where the first European settlers built forts. In 1718, the settlement that was to become New Orleans rose on a natural levee, only to be waterlogged by a high flood. There was nowhere to go, so the settlers raised artificial barriers. At first, the main levee was just under a meter high, raised in part by a law requiring homeowners to do so— not that many of them did. Fortunately for the growing settlement, the floods could spread widely on the eastern bank of the river where there were no artificial levees. In 1727, the French colonial governor proudly announced that the levee was complete. Nevertheless, the city was inundated in 1735 and again in 1785. The intervals between severe floods were long enough that generational memory faded, but the levees were slowly extended along the west bank as far as the Old River, some 322 kilometers upstream and along the east bank as far as Baton Rouge. The levees were far from continuous, mainly protecting plantations. Elsewhere, the flood poured over the countryside as it had always done. Many plantation houses rose on the only higher ground—ancient Indian burial mounds.[4]

The more levees confined the river, the more destructive floods became when they broke through human barriers. By the mid-nineteenth century, it was clear that even raised levees would not contain the river as flows increased. Congress passed the Swamp and Overflow Lands Act in 1850, which deeded millions of hectares of swampland to the states. They in turn sold the swamp to landowners to pay for levees. The new owners drained most of the swamps, converted them into farmland, and then demanded larger and better levees to protect their investments. Catastrophic floods and levee breaks in 1862, 1866, and 1867 finally prompted Congress to form the Mississippi River Commission, charged through the Army Corps of Engineers to "prevent destructive floods."

Figure 13.3 *Repairing a levee on the Lower Mississippi. Drawing by Charles Graham. Author collection.*

The most ruinous inundation of the nineteenth century came in 1882, at which point it was becoming apparent that the Atchafalaya River could become the new main course of the Mississippi. By now the river flowed high above the surrounding landscape, confined only by its levees, a disaster waiting to occur, which finally transpired with the flood of 1927. This caused multiple breaches and destroyed every bridge across the river for hundreds of kilometers. Sixty-seven thousand square kilometers of surrounding country vanished underwater.

The Flood Control Act of 1928 provided funds for coordinated river defenses that were still incomplete in the 1980s. This time, the corps did more than raise levees. They straightened parts of the river, built spillways, like that at Bonnet Carré, nineteen kilometers west of New Orleans, which diverted water away from the city and into Lake Pontchartrain during a major flood in 1937. In places, levees now rose to more than nine meters, but increasingly the Army Corps turned its attention to the Atchafalaya, which flowed into one of the largest river swamps in North America.

Without human interference, the Lower Mississippi's floodwaters would disperse widely over the delta plain through many outlets. Virtually the entire delta would be covered not only with floodwater but also with fine sediment derived from mountains hundreds of kilometers away. As the writer John McPhee remarks, "Southern Louisiana is a very large lump of mountain butter, eight miles thick where it rests upon the continental shelf, half of that under New Orleans, a mile and a third at Red River."[5] Deposits like this compact, condense, and sink. McPhee calls the delta "a superhimalaya upside down." The subsidence continues despite human intervention. Until about 1900, the Mississippi and its tributaries compensated for the subsidence with fresh sediment that came down each year. The delta accumulated unevenly but on the positive side of the geological cash register, as channels shifted and decaying vegetation sank into the flooded silts. The vegetation itself grew as a result of nutrients supplied by the Mississippi.

Before the days of flood defenses, the river spread across the surrounding country quite freely and in many places, except at low water, when it stayed within its natural banks. Today, over three thousand kilometers of levees constrain the river until Baptiste Collette Bayou, ninety-seven kilometers downstream of New Orleans. The delta has lost silt for a century and southern Louisiana is sinking. Meanwhile the river with its levees shoots fine river sediment out into the Gulf of Mexico—some 356,000 tons of it a day. The water rises ever higher behind the levees as the surrounding landscape continues to subside. McPhee calls the delta "an exaggerated Venice, two hundred miles wide—its rivers, its bayous, its artificial canals a trelliswork of water among subsiding lands."[6]

The entire delta is a highly vulnerable, threatened landscape. About half of New Orleans is as much as 4.6 meters below sea level, hedged in between Lake Pontchartrain and the Mississippi. The richest inhabitants live on the highest ground by the river. The poorest dwell at lower elevations in the most vulnerable locations of all. The city receives abundant rainfall, often torrential downpours that cause serious flooding within the river defenses, as happened with Tropical Storm Isaac in 2012. There are no natural outlets, so the water has to be pumped out, which tends to lower the water table and increase subsidence. The pumps, invented by engineer A. Baldwin Wood during the early twentieth century, allowed the city to expand to ever-lower elevations; there was no space else-where.[7] New Orleans is so waterlogged that the dead are buried in cemeteries aboveground. Even from the slightly higher ground of the French Quarter, you look up at the levees and passing ships, their keels well above the city streets.

New Orleans has distinctive levee problems, not least that of the waves caused by the wakes of passing vessels. Everywhere, the high water line is rising as the speed of the confined water from upstream increases and the levees themselves subside. Down at the coast, the shortage of river silt has led to erosion, to the tune of some 130 square kilometers of marshland a year. Louisiana is 4 million square kilometers smaller than it was a century ago. Half a kilometer of marsh reduces the height of a storm surge from offshore by 2.5 centimeters. With the disappearance of 130 square kilometers of coastal barrier, the Mississippi equivalent of the Bangladeshi mangroves is drastically reduced. The vanishing marshes caused the corps to build an encircling levee around New Orleans, turning it into a fortified city—this before one figures in the sea level rises of the future.

At this point, many experts believe that the coast is beyond salvation, as entire parishes vanish and nutrient and sediment starvation wreak

Figure 13.4 *The dedication of A. Baldwin Wood's pumping station in New Orleans, 1915. Author collection.*

havoc on the coastal landscape. Today, much Mississippi water flows down the Atchafalaya, which is the main route by which severe flood-waters reach the Gulf. Morgan City, Louisiana, has a flood stage about 1.2 meters above sea level and lies directly in the flood's path. Huge walls 6.7 meters high surround the small town, which must be one of the most vulnerable human settlements on earth.[8] It is as if the city dwells atop an occasionally active volcano, its fate decided by flood control installations up at Old River, far upstream. This is apart from the storm surges caused by hurricanes. Weather conditions over 42 percent of the United States determine the long-term survival of Morgan City. Floods here last not weeks as they do upstream, but months, for the more people upstream protect themselves with rings of levees, the more water arrives downstream.

THANKS TO HUMAN activity, the Gulf Coast is sinking, sea levels are rising, and humans have effectively reversed nature. Or have they? This is hurricane country, and when such storms arrive they tear into the coast and accelerate erosion dramatically, making for an even more severe threat of destruction from the south than from the north.

Hurricanes entered recorded delta history dramatically in 1722, when one such storm led some French settlers of New Orleans to suggest that the city was uninhabitable. In 1779, another even more severe hurricane razed the city completely. A long history of devastation of barrier islands and inland villages also haunts the coast, but people have always stayed—to farm, to fish, and, in more recent times, because of oil. Rising sea levels compound the problem as they encroach on the eroding and subsiding delta shore. By the 1970s, the estimated annual loss was about a hundred square kilometers a year. The rate has slowed since then, as oil and gas activity has declined somewhat and measures were taken to reduce the loss, which now stands at somewhere between sixty-five and ninety square kilometers annually. Then there are sea surges, which created some seven hundred square kilometers of open water at the expense of wetland in the 2005 hurricane season alone.

In earlier times, before human interference, sediment maintained the

elevation of the wetlands and fed the offshore barrier islands, many of
which are now disappearing in the face of hurricanes and rising sea levels. Tidal gauge data for Louisiana record a sea level rise of nearly a
meter over the past century, most of it caused not by global changes in
ocean volume but by subsidence. If the local sea level were to rise 1.8
meters between now and 2100, all efforts to restore the wetlands
would be canceled out and New Orleans would be seriously threatened.
This is why Hurricane Katrina was a defining moment in the adversarial
relationship between humans and the ocean.

Poverty Point may have been vulnerable to river floods, sea level shifts,
and climate change, but its people could disperse into villages and resettle
elsewhere. Today, millions of people live on the low-lying coastal plains of
the Lower Mississippi region, where warming temperatures, rising sea
levels, and a projected higher frequency of extreme weather events pose
serious threats to much larger, far more vulnerable populations. When
Katrina came ashore on the Gulf Coast in 2005, Americans received a
harsh lesson in the frightening vulnerability of densely populated cities
lurking behind flood defenses.

Hurricane Katrina was only a Category 1 Atlantic hurricane as it
crossed southern Florida.[9] It strengthened dramatically in the Gulf of
Mexico before making landfall in southeastern Louisiana as a Category 3
on August 29, with sustained winds of 195 kilometers an hour. The storm
lost hurricane strength only about 240 kilometers inland, near Merid-
ian, Mississippi. Between 200 and 250 millimeters of rain fell in Louisiana as the hurricane swept inland. The sea surge was far more damaging
than the wind and rain. The height was at least eight meters, inundating
the parishes surrounding Lake Pontchartrain. When combined with
levee breaches, the damage was catastrophic, with significant loss of life,
quite apart from the devastation of coastal wetlands.

By August 31, 80 percent of New Orleans was underwater, in places
4.6 meters deep. A citywide evacuation order came into effect, at first
voluntary, then mandatory, the first such evacuation in the city's history,
despite the ever-present threat of flood. By the time the hurricane came
ashore, over a million people had fled New Orleans and its immediate

suburbs. Nevertheless, over a hundred thousand remained, especially the elderly and poor. An estimated twenty thousand took refuge at the Louisiana Superdome, officially designated as a "place of last resort," and designed to withstand very strong winds indeed. Flooding stranded many residents, many of them on rooftops or trapped in attics. There were bodies floating in the eastern streets; water was undrinkable; power outages were widespread.

The official death toll was nearly fifteen hundred people. As search-and-rescue operations intensified, looting and violence became epidemic through much of the city. Residents simply took food and other essentials from unstaffed grocery stories; armed robberies were commonplace. The authorities imposed a curfew, and then declared a state of emergency as sixty-five hundred National Guardsmen arrived to help restore order.

With flooding and chaos in the city, evacuation seemed a logical strategy. For years, coastal evacuation policies had assumed that most people could afford to leave their houses when ordered to do so and would be able to evacuate in their vehicles. The evacuees would seek shelter with relatives or stay in hotels or motels at higher elevations inland. Katrina proved the planners wrong. More than a quarter of New Orleans residents had no access to an automobile. They lived from one pay period to the next and had no surplus funds for evacuation or other emergencies. This was why many residents sought refuge at the Superdome and convention center. Hurricanes come ashore in this area relatively rarely, which means that generational memories fade quickly. Many people, who had forgotten the experience of Hurricanes Betsy (1965) and Camille (1969), preferred to stay and ride out the storm. Those who stayed were predominantly less educated, poor, and earning lower incomes. These were the people most affected by the disaster.

Most evacuees stayed within three hundred kilometers of New Orleans, but others dispersed all over the country, as far as California, Chicago, and New York. Orleans Parish had a population of 455,188 before the hurricane, and only 343,829 afterward, a drop of 24.5 percent.[10] The hardest-hit parish, St. Bernard near Lake Pontchartrain, lost 45.6 percent of its people to out-migration. Over 250,000 Katrina

migrants went to Houston. How many have returned is a matter of debate, for reliable statistics are hard to develop or come by. Many have
chosen to relocate permanently.

One cannot entirely blame them, given the controversy and faction-
alism surrounding the recovery effort. One reformist faction wanted to
use the storm as an excuse to remake New Orleans in a more efficient,
modern form, including a replacement for the old public school system.
Such a restoration would have required enormous sums of money from
outside. New Orleans lacks a Coca-Cola or other large corporation to
help pay the bill. Congress and the Bush administration did spend large
sums on the city, but unfortunately they rejected proposals for a large-
scale buyout of inundated housing as a basis for redeveloping the entire
city. Instead, Washington gave billions in grants to individual homeowners, who wanted to return, and also for such projects as building schools,
libraries, and water treatment plants. This was all fine and good, but the
money was distributed inefficiently, much of it after very long delays. A
patchwork of redevelopment made it nearly impossible for the poverty-
stricken city government to provide basic services such as garbage collection, fire protection, and policing in a systematic manner.

All of these delays and often mistaken decisions tended to reinforce a
widespread impression that no one outside the city cared about a poor,
largely black population. There was even a sense among black community members that neither the Bush administration nor white New
Orleans wanted blacks to return. As Nicholas Lemann wrote in the
New York Review of Books, an ancient fear of black insurrection tended
to resurface, accompanied by a longing for the city to be reborn as "another Charleston or Savannah, smaller, neater, safer, whiter, and relieved
of the obligation to try to be a significant modern multicultural city."[11]
This was, of course, merely a feeling and something that never even
slightly surfaced as city policy.

Racial undercurrents of all kinds were at the heart of the toxic politics of rebuilding that emerged just as soon as the floodwaters receded.
A Republican real estate developer assigned to the design of a compre-
hensive rebuilding program by the mayor, Ray Nagin, suggested that
some of the most devastated, poorest, and lowest-lying areas of the city

not be rebuilt immediately. Black outrage reached a fever pitch. Many out-migrants felt they were being deliberately excluded from their homes. The chasm between black and white widened even further. This debacle killed the mayor's plan and meant that there was no blueprint at all. As recently as 2010, fifty thousand houses were still empty, more than a quarter of the available housing in the city. The current mayor, Mitch Landrieu, is moving cautiously. He is willing to tear the derelict houses down as part of a new policy that proclaims that every neighborhood will be rebuilt and that one cannot simply wait for displaced residents to return.

Meanwhile, the long-term consequences of the disaster continue. The city is much smaller, so population-based federal funds for housing, health care, and infrastructure are much reduced. Old political districts will be redrawn, resulting inevitably in less black representation at both the state and federal levels and a loss of political power. The permanent displacement of many poorer inhabitants has increased the percentage of educated, high-income residents. The redevelopment of low-income housing is glacially slow, causing some to question whether the recovery effort is equitable. By 2010, about 54 percent of the evacuees had returned to their pre-Katrina addresses. There are powerful reasons to stay away: the lack of affordable housing and a shortage of rental properties. Much recovery funding went to homeowners, who are in a minority among poorer citizens. With little money, few job opportunities, and a shortage of affordable housing, many refugees left permanently.

THE PLIGHT OF the Mississippi delta confirms that we have few options when confronted by extreme weather events made ever more dangerous by rising seas. Armoring cities and coasts with levees and seawalls is, at best, a costly gamble, even if twenty-first-century construction and technology provides ever more secure barriers. Even adopting building codes that allow for surging hurricanes and permit people to take shelter in their homes is an expensive hedge against catastrophic destruction. The only other option is carefully managed relocation at short notice. Katrina and other major storms teach us that the density of population in large twenty-first-century cities is such that our ability to relocate displaced residents either temporarily or permanently is severely limited. Inevitably, situations arise where those who can, leave and those who cannot, stay, fueling already festering economic and social inequities within and between segments of society. The resulting economic and political tensions within communities, between source and host communities, and, in the case of Bangladesh, between nations, do not augur well for a warmer future of higher sea levels and a great incidence of storm surges nurtured by extreme weather events.

"Here the Tide Is Ruled, by the Wind, the Moon and Us"

OCTOBER 11, 1634: "At six o'clock at night the Lord God began to ful-
minate with wind and rain from the east, at seven he turned the wind to
the southwest and let it blow so strongly that hardly any man could walk
or stand . . . The Lord God [sent] thunder, rain, hail, lightning and such
a powerful wind that the Earth's foundation was shaken."[1] Peter Sax of
Koldenbüttel in Schleswig-Holstein was in no doubt of the cause. Divine
wrath in the form of a great sea surge descended on the Strand islands
off northern Germany 272 years after the Grote Mandrenke—or so
people still believed at the time.

Once again, the attack was unrelenting. The people of Strand had rebuilt most of their dikes after the earlier disaster, but their troubles were
not over. The Thirty Years' War brought fighting to the islands, which
resulted in the neglect and inevitable weakening of coastal defenses. In
1625, huge ice floes damaged many dikes; several powerful sea surges
further weakened the inadequately maintained defenses, so much so that
some of them even gave way during summer gales. Then, on October 11,
1634, a strong tempest backed to the southwest, then northwest, coincid-
ing with unusually high tides, violent winds, and destructive swells. A
Dutch hydraulic engineer, Jan Leeghwater, was carrying out land reclamation locally and left an account of "a great storm and bad weather."
His son woke him up in the middle of the night as menacing waves attacked the top of the nearby dike. They fled to a manor house on higher
ground, where thirty-eight people had taken refuge. Raging water from
a nearby tidal canal undermined the foundations and the house fell

Figure 14.1 *Destruction wrought by the Burchardi flood, ca. 1634.*
Victims cling to the roofs of their houses. Author collection.

apart. Leeghwater and his son barely escaped with their lives. Riding along
the shore some days later he saw "many different dead beasts, beams of
houses, smashed wagons . . . many a human body who had drowned . . .
Great sea ships were standing on the dike."[2] The great surge became
known as the second Grote Mandrenke or the Burchardi flood.

Contemporary accounts speak of a storm surge that brought water
levels on the mainland to about four meters above mean high tide. Hundreds of dwellings collapsed or were washed away. Unattended hearths
caused many other abandoned cottages to burn down. As many as
forty-four dike breaks on Strand damaged twenty-one churches and
destroyed thirteen hundred homes and thirty mills. "Wasted are the
Lord's houses," wrote the preacher of Gaikebüll on Nordstrand.[3] Nor
were there preachers to serve them. Saltwater ruined most of the new
harvest. Fifty thousand head of livestock perished. Estimates of human

casualties range between eight thousand and fifteen thousand, at least six thousand people on Strand itself, out of an estimated population of eighty-five hundred. The actual number was probably much higher. Apart from the locals, many migrant workers from elsewhere had been working the land. Their numbers were never recorded.

The long-term consequences were severe, especially on Strand, where many hectares lay below sea level and failed to drain quickly. High tides further widened dike breaches. What had once been a large island was now several much smaller ones. The surviving arable land had to be abandoned because of the unrelenting sea. Thousands of hectares could no longer be used for agriculture. Even where reclamation was possible, recovery was slow, partly because many farmers defied their ruler, Duke Frederick III, and moved to higher ground or to the mainland. Nevertheless, some thirteen thousand hectares lay behind restored dikes on Pellworm Island by 1637. Nearby Nordstrand was another matter. The remaining farmers stubbornly stayed on higher ground and failed to restore their dikes, despite several orders by the duke. Local laws declared that those who did not secure their land against the sea forfeited their right to own it. In the end, the duke expropriated land from the locals and attracted settlers from elsewhere with a charter that promised land and other privileges to people willing to invest in dikes. The incentives effectively gave the newcomers authority over local justice and policing, a remarkable innovation at the time. Several Dutch entrepreneurs like Quirinus Indervelden used Netherlands money to create a new polder in 1654, using expert workers from Brabant Province. Other polders followed in 1657 and 1663. So enduring was the Dutch legacy that a local preacher continued to deliver his sermons in that language until 1870. Today Pellworm and Nordstrand have about nine thousand hectares of reclaimed land. Such was the power of the Burchardi storm surge that it created a tidal channel between the islands that has reached a depth of nearly thirty meters over the past four centuries. (For locations, see figure 5.1.)

Serious dike work along the coast as a whole had begun as early as 1000 C.E. in response to rising sea levels and growing population. However, decentralized administration and the rudimentary technology of the medieval farmer hampered large-scale construction. For five centuries, the fight against the attacking North Sea had been in the hands of major landowners and religious houses, which alone possessed the manpower, organization, and resources to build major dikes.[4] They worked through a morass of local authorities, many of which were inefficient, even lazy. Pumping in particular was virtually nonexistent. For centuries, drainage depended on gravity. Nevertheless, by 1250, most coastal dikes formed a single defensive line. Once this rampart, however weak, was in place, the builders slowly moved the dikes seaward, taking advantage of the sediment built up by years of high and low tides. They would enclose the accumulating deposit, leaving the old dike as a secondary defense. Inside the perimeter, the drained soil consolidated and peat decomposed. As the land subsided, the difference between sea level and the land inshore increased, which was fine until a flood breached the dike or a high tide flowed over the top, in which case the destruction was far greater than before. Moving dikes seaward was sometimes impracticable, especially in delta areas where tides undermined sea defenses. Once this happened, inshore flood works became the primary defense and a great deal of land vanished.

By the early sixteenth century, a growing demand for peat as fuel for expanding cities and for industries such as brewing and pot manufacture led to serious environmental problems.[5] Surface peat was in short supply; water tables were rising. When the peat was worked out, the diggings lay empty and soon filled with water, creating large lakes that threatened nearby dikes and settlements. Many villages and churches fell prey to slowly spreading water that undermined them.

There were occasional sea surges, but the period from about 1530 to 1725 witnessed a slower rise in sea levels, partly as a result of the colder conditions of the Little Ice Age, well illustrated by Dutch artists of the day. The landscape was ripe for reclamation schemes on a larger scale. More centralized hydrological administration helped change the equation. In 1544, King Charles V, ruler of the Burgundian Netherlands, reorganized water control efforts, a process that culminated in the formation of a single water control authority in 1565, which worked closely with polder boards and local communities.

By this time, hydraulic engineering was on the move, thanks to the experience of experts like Andries Vierlingh, dike master to William the Silent, Prince of Orange.[6] Vierlingh spent his life working along the treacherous North Sea coast with its fast-running tides and currents and had also served as a member of several polder boards. His book on reclamation wrote of "roaring breaches." He wrote, "Water will not be compelled by any force, or it will return that force unto you." He added, "One must direct the streams from the shore without vehemence. With subtlety and sweetness you may do much more at low cost."[7] Vierlingh's ambitious ideas coincided with the first widespread use of windmills to pump out polders.

THE NETHERLANDS ARE flat and windy, with an average wind speed of twenty-one kilometers an hour, an ideal environment for wind power. Most likely, the idea of a windmill arrived in the Netherlands in the hands of returning crusaders during the thirteenth century.[8] At first they served as small corn mills, laboriously rotated to face the wind. Primitive windmill pumps lifted water at Alkmaar (today famous for its cheese) as early as 1408, but their scoops could lift water only a mere two meters, more if arranged in series. Over a century later, by 1574, the invention of the movable cap atop the mill allowed just the sails to be rotated, so windmills could be much larger, taller, and more efficient. Wind-driven water pumps led to the development of a new form of polder, created by surrounding shallow bodies of water with an embankment, then pumping them dry. Such polders required capital to build, but paid rich dividends when used for large-scale drainage, even if occasional storms wreaked havoc on embankments. By the later sixteenth century, rich urban merchants had begun to invest in land, as well as to finance large-scale reclamation, which they now considered a potentially lucrative investment as food prices rose and population growth caused shortfalls of good farming land. By 1640, twenty-seven lakes north of Amsterdam had been pumped dry.

Such reclamation works flourished under the direction of Jan Leeghwater of Burchardi storm fame, an accomplished millwright and expert

Figure 14.2 *Windmills at Laandam, Netherlands, 1898. James Batkin (creator).* © *The Print Collector/Heritage/The Image Works.*

on polder reclamation. He would construct an encircling dike and as-sociated canal into which he would pump surplus water with windmills. Once dry, the peat was cut and sold as fuel. The exposed soil then be-came arable land and pasture. In 1607, Leeghwater began draining the 7,000-hectare, 3.5-meter deep Beemster Lake in North Holland. The encircling dike and canal took two years to construct. Twenty-two wind-mills began work before a storm surge destroyed the still-unconsolidated dike. Two years later, dike repairs allowed pumping to resume, this time with forty-two windmills. Trees planted on the flanking dike bound the earth together and provided a pleasant view. Hundreds of men labored with mattocks, spades, and wheelbarrows without any assistance, except from heavy pile drivers operated by teams of thirty to forty workers. Ox-drawn carts, sleds, and large wicker baskets moved clay for the dikes, and also the timber piles and willow wands used for reinforcement. Despite initial setbacks, the Beemster reclamation project was a spectacular success. By 1640, 207 farmsteads, a hundred barns, ten hay or seed stores, a corn mill, two timber wharves, three schools, a church, and a parish hall flourished within the polder. There were even out-of-town houses for city dwellers to enjoy during the summer.[9]

As drainage schemes proliferated, so accurate leveling became ever more important. In 1682, Johannes Hudde, the burgomaster of Amsterdam, organized the first systematic measurement of sea levels, the height being measured by eight large stones, the so-called Amsterdams Peil, the level from which all heights above and below sea level in the Netherlands have been measured ever since.[10]

SYSTEMATIC SEA LEVEL measurements were part of a far wider upgrade of both sea defenses and reclamation. Early sea defenses in the Low Countries were low embankments of tamped-down clay or earth. So were river dikes, which had to withstand not only wave action and tides but also long periods of high water when seepage was a problem. A clay core, gentle landward slopes, and a gravel-filled ditch at the base to catch seepage made them fairly effective.[11]

During the thirteenth century a form of mud dike, the *werdijken*, came into use in northern Holland. The builders piled turves or lumps of sticky clay to form a steep seaward face. A thick seaweed mattress, held in place with piles against the outer face, compressed and rotted into a solid residue that was surprisingly effective against attacking waves. Bricks or stones often reinforced the base, the seaweed piles sometimes extending as high as five meters over the face and top of the embankment. Seaweed clung to itself because of its weight, forming a sweating, elastic cushion that resisted waves. Reeds or wicker mats sufficed where seaweed was unavailable, but were not nearly as durable and had to be replaced every five years or so. Sometimes the builders used individual wooden piles, thirty centimeters square and up to six meters long. Some of these dikes appeared as early as 1440 and remained in use until the nineteenth century.

Sea defenses of earth and timber sufficed until 1731, when the teredo worm, inadvertently brought to the Netherlands by Dutch East India

Company ships, proliferated in Zeeland and elsewhere. This bivalve, which is actually not a worm at all but a saltwater clam, bores into wooden structures such as the hulls of ships immersed in seawater. Teredos ravaged the coastal defenses in short order. With a few years, wooden piles collapsed; seaweed and reed mattresses were washed away. By 1732, fifty kilometers of the Westfriese Zeedijk in North Holland had collapsed. Another twenty kilometers was seriously weakened. Memories of the Christmas flood of 1717 were fresh in everyone's minds, when seventeen thousand people died and water poured into Amsterdam and Haarlem. "The blessed Netherlands . . . is in danger of being flooded because of a rare gnawing of worms," lamented one observer in 1735.[12] The only solution lay in imported stone, an expensive palliative but one offset by much greater durability.

Lengthy trials led to gently sloping stone revetments, walls that extended from the base of the dike to a meter or so above extreme high tide. After 1775, dike revetments consisted of irregular boulders, the cracks between them filled with rubble, the whole resting on a bed of straw, which served to absorb the shock of breaking waves and reduce erosion and seepage. Today's dikes have a sand core, covered by a thick layer of clay to provide waterproofing and erosion resistance. Where there is now land in front of the dike, a layer of crushed rock lies below the waterline and slows water action. A layer of carefully laid basalt rocks or tarmac covers the front surface up to high water level. Grass mantles the rest of the dike, kept dense and short by grazing sheep.

The Industrial Revolution, with its fossil fuels and much more efficient pumps, eventually allowed the Dutch to undertake much larger reclamation projects. A steam pump imported from Looe in southwestern England came into use at Blijdorp near Rotterdam in 1787. At first the farmers opposed its use on the grounds that the noise would scare their cows from giving milk. Opposition soon evaporated when the pump successfully combated winter floods and seepage. Reclamation work had been at virtual standstill in many areas, but now resumed on a larger scale. The greatest impact of steam drainage came at the Haarlemmermeer in the south, which covered nearly eighteen thousand hectares and threatened the major cities nearby. In 1836, a southwesterly gale blew water from the meer to the outskirts of Amsterdam. Another storm menaced Leiden a month later. Steam power now made the draining of the lake possible. An encircling lake and hundred-kilometer-long canal built between 1840 and 1854 allowed three large pumping stations to drain the Haarlemmermeer by 1852.[13]

The nineteenth century also witnessed a frenzy of canal building, aimed at improving access to Amsterdam and Rotterdam for larger oceangoing ships. Sandbanks in the Zuiderzee and circuitous approaches through delta channels made for hazardous, slow passages. Rotterdam was only thirty kilometers from the coast, but the approach by water twisted and turned for a hundred kilometers. Add the delays caused by tides and winds to the equation, and it could take an East Indiaman twenty-one days to reach the open sea and only a hundred more to reach the Indies. The Nieuwe Waterweg to the Rhine mouth was completed in 1872, in the hope that tidal scour would keep it deep, but shallowing began immediately. Only with the advent of steam dredgers did the canal link Rotterdam effectively to the Rhine. By 1876, the twenty-four-kilometer North Sea Canal punched through the coastal dunes and linked Amsterdam to the ocean and, with constant widening and improvements, has served as the main artery for the city's port ever since.

The modern era of sea defenses in the Netherlands can be said to have begun with two defining events—the Zuiderzee flood of 1916 and the flood of 1953. The Zuiderzee flood came on January 14, after several days of rough weather. Water levels rose; winds reached a velocity of over a hundred kilometers an hour. As dikes eroded on both sides, water inundated the island of Marken, where the only protection was low quays. Sixteen people died; fishing boats washed ashore. The material losses were far greater than fatalities, but the disaster stimulated a debate already in progress about reclaiming the Zuiderzee by building a dam across the mouth. On January 13, 1918, the Dutch Parliament passed a bill to start the work. The contractors built the thirty-two-kilometer Afsluitdijk of glacial boulder clay and rocks between 1920 and 1932, with drainage locks that allow excess water to pass into the Waddensee outside. The Zuiderzee became a lake and was renamed the Ijsselmeer. Large polders then reclaimed extensive tracts of valuable farmland from the now freshwater lake.[14]

As the Zuiderzee works continued, a study by the state water authority in 1937 showed that sea defenses in the southwest were an inadequate defense against a major storm surge. The study proposed that all river mouths and sea inlets be dammed permanently in an expensive exercise that would shorten the coast by about four hundred kilometers. World War II intervened and it was not until 1950 that the first stage of dam construction was complete. Then came the second defining catastrophe—the great flood of 1953 that brought disaster to both coasts of the North Sea.

On January 31, 1953, an extratropical cyclone with winds of 145 kilometers per hour developed off northern Scotland.[15] The storm swept into the North Sea. A huge storm surge dashed down Britain's east coast, up the Thames, then across to the Netherlands. Great waves breached dikes and seawalls on the English side in more than twelve hundred places. Forty-six thousand cattle drowned; more than 64,750 hectares of farmland became salt contaminated and useless for many years; over 25,000 houses were damaged or destroyed in villages and towns. Remarkably, only 307 people died, thanks to heroic rescue efforts, but the economic effects were enormous. During the war years, no one had maintained the coastal sea defenses. The storm was a wake-up call. More than fifteen thousand workers, including British and American soldiers with heavy equipment, rushed to repair the damage before the next spring tide in mid-February. Fortunately, London avoided flooding by mere centimeters. The tide level at London Bridge in the heart of the city was nearly two meters above the highest ever recorded. Only a little water flowed over the London embankments. Dike breaks farther downstream had flooded the neighboring countryside and lowered the water level.

The lowest parts of the Netherlands in Zeeland and the southern part of South Holland withered under the onslaught of floodwaters from catastrophic dike breaches. Here, too, the sea defenses suffered from the neglect of the war years, as well as from a sense of security resulting from a lack of serious flooding since 1916. Now the coast paid the price of neglect. Sea levels crested between 3.6 and 4.2 meters above normal and inundated as much as 4,100 square kilometers in short order. As the waters approached, families climbed onto their rooftops, but to no avail. At least 1,800 people died, and 70,000 flood victims required evacuation. Once the sea had flooded Zeeland and most of South Holland, the floodwaters dashed against a major dike, which protected a still-dry Rotterdam and more than three million people in North Holland. Part of the unreinforced dike gave way. The mayor of Nieuwerkerk requisitioned a riverboat and ordered its skipper to close the hole with its hull. Providentially, the barge turned sideward into the gap and acted like a floodgate, saving thousands of hectares from inundation.

When the surge waters finally receded, the destruction was truly catastrophic. In Zeeland alone, the dike breaches were over three kilometers wide. Thirty-nine kilometers of dike were heavily damaged. As recovery efforts got under way with the help of thirty thousand volunteers working on the dikes, the government formed the Delta Committee to make recommendations for the future. Dam construction to close estuary mouths accelerated in the years that followed under a huge project known as the Delta Works.[16] One estuary in the Eastern Scheldt remains open. Following protests from environmental and fisheries groups, the Oosterscheldekering (Eastern Scheldt storm surge barrier) joined an eleven-kilometer gap between two islands, the largest and most ambitious of the thirteen dams that form the Delta Works. Huge sluice-like gates installed in a four-kilometer length of the barrier remain open except when the sea level rises three meters above the regular, specified height, a legally mandated figure. The Oosterscheldekering officially opened on November 5, 1987, exactly 467 years after St. Felix's flood of 1530 inundated a huge tract of land upstream of the barrier. An inscription on the artificial island that forms part of the barrier reads: "Here the tide is ruled, by the wind, the moon and us." The cost was staggering: 2.5 billion euros, well over 3.6 billion dollars.

One final project remained, the construction of a huge storm surge barrier for the Nieuwe Waterweg to protect Rotterdam. Two huge curved

Figure 14.3 *The Eastern Scheldt storm surge barrier. Flickr.com/ Fotografie Siebe Swart.*

steel doors provide protection during surges, remaining submerged out of the way most of the time so that ships can pass. The barrier, opened in 1997, protects a million people from inundation.

When completed, the Delta Works reduced coastal sea defense levees by seven hundred kilometers and allowed much better regulation of water for agricultural purposes. The sea defense works are a delicate balance between safety, economic factors, and the environment, but with no compromise on protection. But there is far more to coastal protection than just seawalls and levees, the latter tending to fence in rivers and cause them to flow more rapidly. The Scheldt River provides an example of what the future holds.

The Scheldt estuary has caused political controversy between the Netherlands and Flanders since the sixteenth century, mainly over competition between Antwerp and Rotterdam. The Scheldt Estuary Development Project (ProSes), set up in 2002, aims at fostering more sustainable development throughout the Scheldt estuary in the light of ever-changing sea defenses and future environmental changes to the river further in-

land. The Scheldt estuary is one of the few remaining European river
mouths that encompasses everything from fresh- to saltwater tidal areas
and includes the largest brackish marshes in western Europe.[17] The
challenges of planning for the future are extremely complex, for there
are many stakeholders. A river channel of at least 13.1 meters deep for
the enormous container ships of the future is needed to keep Antwerp's
port competitive, located as it is at the hub of rail and highway networks to the heart of Europe. Farmers are reluctant to give up valuable
hectares for flood basins in an area where agricultural land is in short
supply. Important commercial and recreational fisheries need protection,
while recreational boaters have a right to sheltered water away from
shipping lanes. Each group of stakeholders in the region has valid needs
and concerns, quite apart from the pressing urgencies of nature conservation in an estuary where the area of shallows and saltwater marshes
has shrunk dramatically in recent centuries as a result of land reclamation and urban development. Above all, there is the issue of safety, a
need for protection from rising sea levels and the catastrophic sea surges
that coincide with strong northwesterly storms and high tides. The 1953
disaster was the catalyst for a new chapter in flood control work throughout the Low Countries. In Flanders, another major flood in 1976 that
inundated parts of Antwerp and Ghent also prompted long-term planning for flood protection as far into the future as 2030.

As far as Flanders is concerned, the process of developing their long-
term plans, known as a whole as the Sigma Plan, was, and still is, thwart
with tension, especially over proposals to transform agricultural land into
artificial floodplain as part of a sophisticated plans for controlled flood
management. Some of the flood areas require the compulsory purchase
of six hundred hectares or more of cropland by the government from the
farmers. Belgium suffers from a shortage of open space for recreation,
so the flood control areas are being created not only as potentially inundated basins but also as nature preserves and recreational areas with
hiking trails and bike paths. The engineers chose the flood areas with
great care, then lowered the height of the riverside levees, thereby allowing water generated by very high tides or storm surges to flow over them
into the low-lying area behind them. A much higher dike at the inshore

extremity of the flood basin reaches a height of seven meters or so, while drains in the riverside levees allow floodwater to escape when the tide falls. The entire project not only creates flood control basins but also restores estuary environments that will thrive into the long-term future, when local populations are considerably higher. At present, flood control basins can provide protection against 70- to 180-year storm surges, but the long-term planning continues. Every aspect of the construction and completed projects is subject to constant monitoring of soil nutrients and water quality, as well as changing vegetation and animal populations. So tight is the control that even the sand and mud raised by ship channel dredgers is dumped in carefully specified areas to prevent damage to marshlands.

The Sigma Plan combines coastal protection with attempts to steer and store storm surge and tidal waters in carefully controlled ways, while at the same time paying careful attention to commercial, environmental, and human needs. All of this requires enormous sums and a long-term commitment to flood and coastal protection that worries as much about the development of Antwerp's waterfront as it does about ports, fisheries, and farmlands. Everything looks far into the future; construction and monitoring continues steadily, with no illusions as to the difficulty of balancing modern industrial needs against those of people and the environment. Such approaches, expensive and long-term as they are, offer hope not only to those who live by the banks of the Scheldt, but also in other estuary lands, which may be able to adopt some of the lessons learned—if the funding can be found to help them achieve results.

WHATEVER THE SUCCESSES inland, one can never escape the unpredictable future. Today, the Netherlands sea defenses are the strongest they have ever been, but the work never stops, as new calculations reveal hitherto unsuspected weak spots. The defenses are continually being strengthened and raised to meet a huge safety margin, equivalent to that of a flood on a scale that would occur only once every ten thousand years for the western, and most densely populated, parts of the country. A criterion for floods that occur once every four thousand years applies to less populated areas. Every five years, the authorities test the primary flood defenses against these norms. The results can be sobering. In 2010, about 800 kilometers of a total dike length of 3,500 kilometers failed to meet the norm, which is becoming ever stricter as research on wave action and sea level rise continues.

Have the Dutch mastered rising sea levels? The specter of humanly caused global warming looms. In 2008, the Dutch State Committee for Durable Coast Development, established in response to the Hurricane Katrina disaster in the United States, attempted to answer a fundamental question: Could people continue to live in the coastal zones under the most extreme scenarios? For example, they estimated large sea level rises of 1.3 meters by 2100 and 4 meters by 2200, much higher than other estimates. The committee members recommended measures that would cost about a billion euros (over $1.4 billion) a year. They urged a tenfold increase in safety norms, the strengthening of dikes, and intensified sand replacement to build up coastal beaches. In addition, they suggested lakes in the southwest be used as river water retention basins, at the same time raising the water level in the Ijsselmeer to provide additional freshwater. In response, the government set up a Delta Fund and a Delta Programme that provides not only long-term funding but also a mechanism for individuals, communities, and special interest groups to be involved in an integrated approach that addresses both local and national issues. At the same time, many cities and towns are moving ahead with their own plans as part of the scheme. Rotterdam is trying to make itself sea-surge-proof by 2025, including provisions for a neighborhood of environmentally friendly houseboats, a prototype for the long-term future.

Most coastal villages in the Netherlands have no view of the ocean. The sea hides behind great ramparts that are as high as thirteen meters and up to forty-six meters deep. A steep climb to the top lets you see the North Sea crashing against reinforced concrete and stone. Look behind you and you realize at once just how vulnerable coastal cities, towns, and villages are in a densely populated country of which over half is below sea level. Widening and raising dikes will be an uphill battle against the North Sea, but there is no debate about the need to prepare for the long-term effects of the rising sea levels, increased storminess, and surging rivers predicted for the future. Large-scale engineering is a way of life in the Netherlands. Every year the Dutch spend billions of dollars on infrastructure and sea defenses, on raised seawalls and river dikes, and also on heightened quays in major ports. They invest in massive water barriers and ever-larger pumping stations—all to keep the nation's feet dry.

Defending the Netherlands from the North Sea requires long-term thinking, enormous sums of money that have to be planned for in advance, and a forward-looking strategy for the distant future that allows for more rapid climatic change than may in fact occur. Every plan, every study, every investment has one purpose—long-term security for future generations. There is no room for skepticism about global warming or future climate change along a coastline where 60 percent of the country is below sea level. Fortunately, the Netherlands can afford such long-term thinking. However, the question remains. Can the government and its people really ensure long-term safety? There is talk of planning for 125,000-year norms. But, as observers have pointed out, the 125,000-year event could occur tomorrow, without warning, and one has to be ready for it.

The Dutch and Belgians are sober, dispassionate planners and engineers, who are looking for long-term solutions. They are fortunate to have the resources to develop state-of-the-art armor for their coasts and estuaries. However, the scale of engineering and infrastructure needed for long-term success is so gargantuan that it may defeat us. Can we really rule the tides? The Dutch seem to think so, but only the distant future will prove them right or wrong. In the case of the wealthy, advanced Low Countries, the chances of success are as good as they can get. The same cannot be said of other parts of the world where millions of people live at the mercy of the ocean.

Epilogue

THE ATTACKING OCEAN HAS TAKEN US on a journey back to the Ice Age, to a time when sea levels were as much as 122 meters lower than today. We've traveled through the subsequent nine thousand years of warming that brought global shorelines to their near-modern configurations about a millennium before the first civilizations developed in Mesopotamia and along the Nile. For the next six millennia, the climb was effectively minimal, as the Egyptian pharaohs built the pyramids, complex states developed in South Asia and China, and Rome dominated much of the Western world. The geology may have been relatively quiescent, but human vulnerability to the attacking sea increased dramatically. Coastal populations in low-lying environments like the Nile delta and the great estuaries of what are now Bangladesh, China, and Vietnam rose dramatically from a few tens of thousands to millions. Mushrooming cities acquired a growing vulnerability to natural cataclysms and extreme weather events like tropical cyclones with their ferocious storm surges. The major threats from the ocean were not those of rising shorelines, but from earthquake-caused tsunamis and violent storms, phenomena whose potential for damage ashore increased significantly if local sea levels rose even slightly or subsidence allowed the ocean to surge inshore, contaminating water supplies and flooding agricultural land. There were certainly casualties and suffering, but nothing on the scale that lurks in the foreseeable future, thanks to the sustained warming that began during the height of the Industrial Revolution around 1860. Since then, sea levels have resumed an inexorable climb, which now seems to be accelerating, whence the thoughtless media hysteria that focuses, laser-like, on perceived imminent catastrophe, on a world of rapidly melting ice sheets smothered by seawater. Reality is, of course, much more complex.

History allows us to take a more measured look at the intricate relationship between warming temperatures and rising sea levels. The rapid warm-up immediately after the Ice Age far exceeded the pace of today's changes. But even then sea level rise, while faster than today, was *cumulative*, a matter of fractions of a centimeter, or at the most a centimeter or so a year, sometimes not even that. Sea level rise is a seemingly long-term problem, which is why many people discount the threat. One could do this with some impunity in earlier times, but not today, when tens of millions of us live a few meters above sea level, even below it. In the short term, the greatest threat comes not from cumulative sea level change but from extreme natural events, whether earthquakes, tsunamis, or tropical storms, which spread water *horizontally* over low-lying coastal landscapes and river deltas, some of the most densely inhabited environments on earth. Combine this with tiny annual sea level rises and you have a volatile and growing recipe for long- and short-term disaster; the former is global, the later predominantly local.

MUCH OF THIS book is a litany of destruction wrought by what are, by historical standards, transitory events. But how do these past catastrophes stack up against the challenges of the longer-term future, the next century or so? There are, of course, some low-lying settlements and island populations that are already being forced to move, but such relatively small-scale shifts pale beside the decade- and century-long disruptions that lie ahead in densely populated nations like China, Europe, and the United States.

At a general level, the figures are daunting. Roughly two hundred million people globally live along coastlines less than five meters above today's sea level. By the end of the twenty-first century, this figure is projected to increase to four hundred million to five hundred million. At the same time, coastal megacities will continue their breakneck growth. In Europe, a sea level rise of about a meter will threaten thirteen million people. A billion people live within twenty meters of mean sea level on land measuring only some eight million square kilometers, an area roughly equivalent to that of Brazil. The land loss will affect the gross national product of flooded areas, impinge on expanding urban settlement, inundate agricultural land, reduce job opportunities, and eradicate coastal wetlands that offer a measure of protection against flooding.

Which nations are most vulnerable? Bangladesh is close to the top of any list, as are the Pacific Islands described in chapter 12, and also the Bahamas. Vietnam with its Mekong delta is under threat. More than a third of the river's delta will vanish underwater with a one-meter sea level rise. In a perhaps prophetic event, the Vietnamese government evacuated 350,000 people in the face of Tropical Storm Ketsana in September 2009. The same storm left 80 percent of Manila in the Philippines underwater. Shanghai with its estuary and dense urban population is also high on the list. In Europe, the Low Countries and the southern Baltic coast, including the Oder and Vistula estuaries, are potential victims, as is eastern England. In the Mediterranean, densely populated flatlands such as the Nile delta, and, of course, the Po delta and Venice, are under attack. Without sea level defenses already in place, constructed at vast cost, some of these areas would already be underwater.

What about the United States, where millions of people live at, or close to, today's sea level? Using government elevation databases, researchers at the University of Arizona have analyzed the vulnerability of every coastal city in the lower forty-eight states with a population of over 50,000 people.[1] The results provide sobering food for thought. Coastal cities along the Gulf and southern Atlantic coasts will be especially hard hit. Miami, New Orleans, Tampa, and Virginia Beach, Virginia, could lose more than 10 percent of their land areas by 2100. An average of 9 percent of land within 180 US coastal cities could be threatened within the same time frame. Collectively 40.5 million people live in these cities, twenty of which have populations of over 200,000. This is apart from erosion and resulting temporary flooding as well as storm damage in a future of more extreme events.

Global warming has raised sea levels about 20 centimeters since 1880 and the rate of rise is accelerating. Many scientists expect a rise of 20 to 203 centimeters this century, depending on the release of greenhouse gases and other pollutants into the atmosphere. More specifically, the Arizona study projects a 2.5- to 20-centimeter climb by 2030 and a 10- to 49-centimeter rise by 2050, the amount varying considerably from one location to another. If current rates of greenhouse gas emissions continue, global temperatures will rise to an average of thirteen degrees Celsius warmer than today by 2100. According to Jeremy Weiss of the University of Arizona, this would lock us into at least 4 to 6 meters of sea level rise in subsequent centuries, as parts of the Greenland and Antarctic ice sheets dissolve.[2] With an almost 3-meter rise, nine large cities, including Boston and New York, will have lost 10 percent of their current land areas. With a near 6-meter rise, about a third of the land area within US coastal cities will have vanished.

Apart from sea level rises flooding urban acreage, there is the additional issue of vulnerability to so-called hundred-year storms. More than two thirds of the studied locations now have a more than doubled risk of a hundred-year tempest within the next eighteen years. The figure is even higher for areas outside the Gulf of Mexico. These flood levels are often 1.2 meters above average high tide levels, yet across the country nearly five million people live in houses less than 1.2 meters above high tide. In 285 cities and towns, more than half the population lives on land below this line. Nearly four million people dwell less than a meter above high tide. About half of them reside in Florida, where eight out of ten of the most threatened cities lie. About thirty billion dollars in taxable property is vulnerable below the 1.2-meter line in just three coastal counties in southeastern Florida alone—quite apart from Miami-Dade, the most threatened county in the state.

None of these figures take into consideration the potential damage that could be caused by a mere 1.2 meters of flooding to dry land devoid of housing, including 7,770,00 hectares of roads, bridges, commercial buildings, military bases, airports, and agricultural land—and the list goes on and on.

INEVITABLY, THERE ARE those who argue that the risks are negligible at this stage, that we shouldn't concern ourselves with such long-term threats. They accuse the experts of relying on inadequate data, on faulty computer models and too few observation points along the coast. One can only point out that the data is improving literally monthly and is the best available. Science is, after all, a cumulative progress. Those who deny the reality of global warming and accompanying sea level rises are flogging a dead horse. Many of them have well-scripted agendas or a blind faith in some form of usually outdated political ideology. We're now at a point where we can no longer behave like ostriches with our heads in the sand.

Consider the climatic projections for California, produced by an impressive army of highly qualified experts.[3] Thoroughly researched projections speak of a 60 percent reduction in mountain snowpack by the year 2100, the snowfall that yields much of the state's water supplies. Heat waves and severe storms will become more commonplace and the Pacific along the California shoreline may rise as much as 122 centimeters within a century. Future sea levels will be higher here than in Oregon or Washington, for the movements of tectonic plates are causing the coast to sink as higher temperatures and melting ice contribute simultaneously to a rising Pacific.

The effects of this climb are already with us. Reliable measurements from the Golden Gate document a rise of eighteen centimeters over the past century. The projected sea level rise above this eighteen centimeters may seem like an academic figure, but most emphatically not for an enclosed bay, where much of the surrounding shoreline lies only a few meters above sea level and intensive groundwater pumping has caused widespread subsidence. At issue here is not so much the actual higher water level, but a future in which a higher incidence of major winter storms will coincide with high tides, causing the sea to cascade much farther inshore and to remain longer on the flooded land. Imagine a future when the runways at Oakland and San Francisco Airports remain partially underwater for weeks, even months, or when storm surges inundate freeways close to high tide level. The disruptions would be enormous, to say nothing of the flooding and destruction of buildings close to sea level. This future is reality, not science fiction.

In the San Francisco Bay area and elsewhere, sea level changes over the past century, and especially over the past two decades, have created what can be called a launchpad for storms and exceptionally high tides, somewhat akin to the effects of water overflowing from a shallow lake in flat terrain that spreads over surprisingly long distances. History tells us that even modest sea level climbs increase such storm-related flooding dramatically. Even many centuries ago with many fewer people around, these events led to thousands of casualties, famine, and even the collapse of royal dynasties. The threat is infinitely higher and more urgent today, not only because of storms and warming, but also because of a second reality: the enormous numbers of people at risk along the world's coasts—at least two hundred million of us and climbing.

With more severe storms and extreme weather events projected for the future, even a few centimeters make a profound difference between the once-a-decade flood and a hundred-year storm surge. Of course, flood levels will reach different levels in individual locations and on diverse timelines. For example, the US Gulf Coast experiences more major storms and accompanying storm surges than other portions of the North American shoreline. Hurricane Isaac came ashore near New Orleans in August 2012 and brought widespread flooding. Katrina destroyed levees and caused catastrophic loss of life. A recent national study of flood-threatened areas found that over half the sites examined had a one-in-two chance of water reaching a level 1.2 meters higher than the average local high tide by 2030.[4] By 2050, many locations should experience 1.5-meter or higher floods above high tide on a regular basis. Beach homes and mansions line kilometer after kilometer of low-lying coastline, inviting destruction. Perhaps it's also worth mentioning that current sea level studies assume that recent historical storm patterns won't change. However, in a warming future, the frequency of storm surges and other extreme weather events could affect the extent of coastal flooding in dramatic ways.

WHAT'S TO BE DONE? No one has entirely satisfying or utterly realistic solutions. The options are limited and almost invariably dauntingly expensive. Do we build enormous sea defenses and wall off the ocean, as the Netherlands have been doing for centuries? Over many centuries, the Dutch have created formidable armor designed to combat the kinds of storms and sea surges that blow ashore perhaps only once in a millennium or perhaps only every hundred thousand years—but no one knows when the epochal storm surge will come. There are numerous other examples. Saint Petersburg in Russia was built on a swamp, fed by the Neva River and the Baltic Sea. Twenty-six kilometers of levees and gates shield the city from floodwater. A massive seawall erected at vast expense with Japanese funding protects Malé, the capital of the Maldive Islands in the Indian Ocean, which is only a few meters above the ocean.

Could one develop a sea defense system to protect the New York harbor region and the extremely valuable real estate surrounding it? (It's worth noting that the original Dutch settlers built their tiny community in south Manhattan on higher ground, safe from flooding. Only much later, after the British takeover in 1664, and as New York became an increasingly important shipping center, did settlement expand into lower-lying areas. Perhaps the Dutch remembered their experience at home.) The cost of building New York a sea defense system would be at least $15 billion. And then, as many experts have pointed out, you have to maintain and upgrade them over the long term, and that is very expensive as well. The city has already taken modest steps such as installing floodgates at sewage plants and raising the ground level while developing low-lying areas in Queens. At the same time, subway design and other infrastructure projects incorporate the best guesses and estimates of the time as to what would be the worst that nature could throw at the city as 100- or 500-year events, but the increasing frequency of severe storms on an almost annual basis creates a new dynamic.

"Anyone who says there's not a dramatic change in weather patterns, I think is denying reality," said New York governor Andrew Cuomo at a press conference on October 30, 2012. New York mayor Mike Bloomberg agreed, adding, "What is clear is that the storms that we've experienced in the last year or so, around this country and around the world, are much more severe than before."[5] The levee solution has been widely used in the New Orleans region, but as experience has shown, such barriers are fallible unless properly built and, again, that is expensive. Levees are in place around Hoboken, New Jersey, but the Hudson River rose when Sandy came and swept over the levees at both ends of the city, flooding some lower-lying neighborhoods and turning others into islands. Over twenty thousand people had to be rescued from their homes. The waters were a brew of rainwater, river water, and sewage, ponded like water in a bathtub inside the levees. There was only a single pumping station to drain the streets.

In New York's case, the only solution may be to wall off the city with a levee-like barrier that would halt a nine-meter surge, twice the height of Sandy's. Movable gates would shut the city off from the Atlantic when

Figure 15.1 *Hurricane Sandy comes ashore in Southampton, New York. Lucas Jackson/Reuters.*

Figure 15.2 *Destruction and fire damage caused by Hurricane Sandy in Breezy Point, Queens, New York. Peter Foley/EPA/Newscom.*

necessary, but would allow ships, river water, and tides to flow freely most of the time. An attractive proposition, perhaps, given the enormous cost of Sandy's destruction, but it bristles with environmental difficulties. How would the barrier affect tidal flows and currents? What about sediment buildup and other environmental problems? Then there are thorny social issues. Would the protection be socially equitable? To put it bluntly, who would be behind the barrier and who would not? All this is quite apart from a longer-term question. Would the barrier work given the uncertainties of climatic change and the growing magnitude of extreme weather events? In any case, for many areas like Rockaways and Coney Island on the Long Island shore, the only feasible solutions would be seawalls or moving everything to higher ground.

Whatever the proposals for New York are adopted, they will be a long time in being implemented, even if the politicians support them. These are long-term projects that extend far beyond our generations and those of our children, which is a hard concept for politicians obsessed with election cycles to grasp. Permits and environmental studies alone will also take years. However, Sandy gives us notice that the federal and state governments involved would be wise to start specific, long-term planning immediately, even if the payoff is beyond our lifetime. Expensive they may be, but they can be cost effective. One can only cite the example of Providence, Rhode Island, which protected its low-lying areas with a nine-hundred-meter gated barrier as long ago as 1966, after two destructive hurricanes in 1938 and 1954. The barrier prevented flooding during Sandy and on earlier occasions. Most of the engineering for this and other such projects is relatively straightforward. It is a matter of looking far into the future and having the initiative and courage to protect us now.

Some of the country's most urgent needs are fairly immediate—we need to improve satellite coverage for weather forecasting; we need to make massive long-term investments in smart grid technology, especially important for a nation with an outdated infrastructure that is heavily dependent on electrical power; and we need to maintain a strong federal presence in emergency response, a lesson that came through strongly after Sandy, an enormous storm whose effects extended far beyond the boundaries of a single state. Above all, governments have to assume that warmer oceans will breed stronger storms and plan accordingly. While we cut carbon emissions as the only long-term palliative strategy, we will have to use both resilience strategies and massive infrastructure improvements to live in what is becoming an increasingly hot and more crowded world.

WHATEVER WE DO is going to be expensive. Armoring the shore anywhere is hideously costly and requires a combination of long-term political will, strong public support, and very deep pockets indeed. There are no guarantees seawalls will work in the very long term. Costly sea defenses on the scale of the Netherlands are impracticable for poor countries like Bangladesh, where, in any case, the geology and local opposition militate against such construction. The problem is left to local improvisation and ingenuity, which is probably insufficient when millions of people are involved. Tidal barriers such as those that protect

London, New Orleans, and those proposed for other floodplain cities such as Shanghai are another expensive palliative. Again, cost is a deciding factor, just as with dikes and levees.

Low-lying coastal areas are favorites for developers, who cater to people's desire to live by the ocean, to see the breakers. Many of the world's most important ports also lie in low estuaries, where thousands of hectares of mangrove swamps and wetlands once thrived. Bulldozers have swept away about the most effective natural protection against the rising sea available to us. Only recently have we realized just how effective such seemingly useless marshlands are in protecting us from sea surges and rising seas, especially from the full effects of hurricanes, where mangroves can protect entire communities from destruction, a lesson learned from Hurricane Andrew in Florida in 1992. We've stripped away most of nature's natural armor against the sea in a promiscuous quest for prosperity, perhaps the wrong kind of prosperity. Restoring wetlands is an attractive palliative, but it is probably too late, given the intense development of many coastal areas. We face a future that we are not prepared to handle, and it's questionable just how much most of us think about it.

There are those who choose to ignore the attacking sea by pushing the challenge onto future generations. In China, despite chronic problems with rising sea levels, storm surges, and subsidence caused by human activity, the Shanghai authorities persist in filling in low-lying coastline to allow further residential and industrial development. They are choosing to invest in seawalls and flood control gates. In Florida, high rise construction along threatened beachfronts continues apace, with the state spending billions of dollars on coastal protection such as seawalls and breakwaters that are, in the long term, but temporary palliatives. One can only describe the construction of artificial islands by the government of Dubai in the Gulf region for thousands of new houses at or near sea level as what two respected geoscientists have called "a stunning act of delusional hubris."[6]

If armor is too expensive and wetlands are gone, what options remain? They are a matter of adaptation and ingenuity, almost all of them at the local level: Require developers to elevate buildings, then protect

their foundations with seawalls; more radically, create floating neighborhoods, where entire tracts float on pontoons in sheltered waters that
rise and fall with tides, surges, and changing sea levels. Experiments
with floating houses (not to be confused with houseboats) are under way
in the Low Countries and may offer a viable, reasonably affordable solution. Given that many Bangladeshi farmers spend weeks a year effectively afloat in partially waterlogged dwellings or in boats, this may offer
one low-cost solution, provided there is enough farming land or other
economic activity above sea level—and protection against cyclone-driven
sea surges.

Finally, there's the option of managed retreat. As we've seen in earlier
chapters, this has been the strategy of choice for hundreds of thousands
of years, an option easily exercised when hundreds or thousands, rather
than millions, of people are involved. This is no longer possible, for we
live in a hemmed-in world of coastal megacities, policed national frontiers and densely packed rural populations. We can no longer move
freely to higher ground. The rules of managed retreat have changed beyond recognition. We're approaching a time, perhaps within half a century, when climate change and sea level rise will force millions of people
to move to higher ground, to become climatic refugees. We can no longer kick the problem downstream for our grandchildren. We're looking
at a future of accelerating humanitarian crises that may involve reset-
tling millions of people in completely different rural and urban environments.[7] Obviously, some of the migrants will have the resources to cope
on their own, but most will be people with no means of moving in
planned, safe ways, creating daunting challenges for the global community. Just developing employment opportunities and infrastructure as
well as housing and water supplies and integrating thousands of migrants into entirely different cultural traditions will become pressing
international problems.

In coming decades, we'll have to face daunting questions. Are there
ways in which one can help people adapt to the loss of agricultural land,
other than by forced movement of thousands, even millions, of people
to higher ground? How can governments and nongovernmental organizations handle massive population shifts from one overburdened city to

an unwilling neighbor or from one crowded rural area to where land is already in short supply? Such movements, triggered by loss of sustainability, water shortages, and so on, are a traditional response for rural farmers, especially if they have relatives dwelling in landscapes elsewhere. We would do well to remember the lesson of history—that people move in search of food when confronted with hunger. Random movements of thousands of starving victims marked the great Chinese and Indian famines of the late nineteenth century, and an ancient Egypt threatened by drought as early as 2180 B.C.E.[8] We would do well to remember that land losses from sea level rise are permanent, unlike those from storm surges or tsunamis. The land loss is enduring, the damage to groundwater usually irreversible.

How does one build long-term resilience to environmental changes like rising sea levels? The answer lies in levels of international cooperation and funding to handle migration unheard of in today's world. Our descendants will have to address migration in a much broader context than just that of environmental threats, recognizing the complexity of all human movements. Above all, they'd have to accommodate a future where significant numbers of people living in low-lying environments could be trapped unless mechanisms exist for them to adapt—and if necessary, move elsewhere. With tens of millions of people involved and a timescale of a half century, at the most two centuries, the problem of climatic migration demands international attention now, not in our great-grandchildren's time. To bury our heads in the proverbial sand in denial is not an option. Ten millennia ago, the fisherfolk of Doggerland could adjust effortlessly to their changing world. Many of us, and certainly not our descendants, will never have the option of free movement in the face of the attacking sea. There are too many of us on earth. We face a frightening, long-term crisis that will challenge all our deep reservoirs of ingenuity and opportunism. The sooner we confront our predicament head-on, the better, for our challenge is to master the earth.

Acknowledgments

The Attacking Ocean began in the middle of a lecture on the sea level challenges faced by Bangladesh delivered by Major General A. N. M. Muniruzzaman (retired) at an Aspen Environment Forum some years ago. His presentation not only shook me, but compelled me to write this book. I am grateful to him for his inspiration. In the end, the book turned into something very different from what I had originally imagined, particularly with its emphasis on extreme events of all kinds, which have affected human societies since the Ice Age—and with even greater force today. From the beginning, I chose to write a global book, which describes ancient (and often modern) societies from all corners of the world. In the final analysis, I am convinced this was the right approach, given the remarkable levels of ignorance about rising sea levels on the part of a public nurtured on sensationalist headlines about global warming, melting ice sheets, and scenarios that conjure images of the impending inundation of London, New York, and places east and west, if not by rising sea levels, then by hurricanes and other extreme unpleasantnesses. I hope this book offers a more sober assessment.

This is a book about environmental vulnerability, about rising populations and various ways of ameliorating the effects of rising sea levels. Above all, however, it's a story not about potential technological solutions, but about people and the ways in which they live with the ocean and will live with it in the future. I've attempted to write a story that is very much my take on a confusing jigsaw puzzle of archaeology, geology, history, and paleoclimatology, with excursions into subjects as diverse, and sometimes esoteric, as barrier islands, mangrove swamps, and nineteenth-century accounts of tropical cyclones. I am, of course, responsible for the conclusions and accuracy of this book, and, no doubt, will hear in short order from those kind, often anonymous, individuals, who delight in pointing out errors large and small. Let me thank them in advance.

This book is the culmination of years of research into ancient climate change. It draws on a vast academic literature, much of which is virtually unknown to nonspecialists. I relied not only on archives and libraries, but also on years of discussions with colleagues from many disciplines either in person or across cyberspace. I realize now that the story took so long to mature in my mind, much of it years before the esteemed major general's talk, that I cannot possibly remember you all. Please forgive me if I offer a collective thank-you. You have my eternal gratitude. Special thanks to: the late professor Grahame Clark, Nadia Durrani, Vince Gaffney, Chris Jacobson, Paul Mayewski, George Michaels, Stefaan Nollet, Annet Nieuwhof, Bruce Parker, Helga Vanderveken, and Stephanie Wynne-Jones. My thanks also to those many folk who have asked perceptive questions at lectures or in seminars. They have helped me immeasurably.

As always, Peter Ginna and Pete Beatty have made this book infinitely better. It is, as I have said of earlier works, as much theirs as mine. Their perceptive comments and suggestions were invaluable. I enjoy my relationship with them more than I can say. My old friend Shelly Lowenkopf talked me through difficult moments and was, as always, a companionable and astute critic in a partnership that goes back many years. So does my friendship with Steve Brown, who drew the maps with his usual skill. Susan Rabiner is the best of agents and always provided timely, and total, support. She is always there for me. And lastly, my gratitude, again as always, to Lesley and Ana, who provide encouragement and laughs at the right moments, and to our cats (once kittens), who have found that the fastest way to the great outside is across my desk and its tempting keyboard.

Notes

Chapter 1   Minus One Hundred Twenty-Two Meters
and Climbing

(1) A layperson's account: Mark Maslin, "The Climatic Rollercoaster" in *The Complete Ice Age*, ed. Brian Fagan (London and New York: Thames and Hudson, 2009), 62–91.

(2) Brian Fagan, *Beyond the Blue Horizon* (New York: Bloomsbury Press, 2012), chap. 2.

(3) Eustacy and isostasy are well described by Orrin H. Pilkey and Rob Young, *The Rising Sea* (Washington, DC: Island Press, 2009), 31ff.

(4)   Pilkey and Young, *Rising Sea*, 35ff.

(5) Pilkey and Young, *Rising Sea*, 79.

(6) Bruce Parker, *The Power of the Sea* (New York: Palgrave Macmillan, 2010), 212ff, discusses rising sea levels.

(7)   The TOPEX/Poseidon satellite, which functioned between 1992 and 2006, was a joint project by French and American scientists. The satellite measured the surface topography and height of the oceans.

(8) Nicholas Shrady, *The Last Day: Wrath, Ruin, and Reason in the Great Lisbon Earthquake of 1755* (New York: Viking, 2008). Parker, *The Power,* 133ff, also has a vivid description. See also the eyewitness accounts by various writers who were in affected places at the time of the earthquake, published in the *Philosophical Transactions of the Royal Society* 49 (1756): 398–444.

(9) Quoted by Parker, *The Power,* 134.

(10) Parker, *The Power,* 135.

(11) Parker, *The Power,* 136–42, has an excellent description of tsunamis.

(12) Stein Bondevik et al., "Record-Breaking Height for 6,000-Year-Old Tsunami in the North Atlantic," *EOS* 84, no. 31 (2003): 298, 293.

(13) A huge literature surrounds this epochal disaster. The most accessible

description of Akrotiri: Christos Doumas, *Thera: Pompeii of the Aegean* (London: Thames and Hudson, 1983).

(14) Plato, and Desmond Lee, trans., *Timaeus and Critias* (London: SMK Books, 2010), 25c–d.

(15) Thucydides, and Richard Crawley, trans., *History of the Peloponnesian War* (Toronto: University of Toronto Press, 2011), book 3:89: 2–5. Accessed at http://classics.mit.edu/Thucydides/pelopwar.3.third.html.

(16) Simon Winchester, *Krakatoa: The Day the World Exploded: August 27, 1883* (New York: HarperCollins, 2003).

Millennia of Dramatic Change

(1) I'm grateful to Professor Peter Rowly-Conwy of Durham University for his advice on this point.

Chapter 2   Doggerland

(1) M. C. Burkitt, "Maglemose Harpoon Dredged Up from the North Sea," *Man* 239 (1932): 96–102. See also J. G. D. Clark, *The Mesolithic Age in Britain* (Cambridge: Cambridge University Press, 1932), 125.

(2) This chapter draws heavily on Vincent Gaffney, S. Fitch, and D. Smith, *Europe's Lost World: The Rediscovery of Doggerland* (York: Council for British Archaeology, 2009).

(3) For an analysis of Reid's work, see Gaffney et al., *Europe's Lost World*, 3–13.

(4) Quoted from Gaffney et al., *Europe's Lost World*, 3.

(5) Clement Reid, *Submerged Forests* (Oxford: Cambridge Series of Manuals of Literature and Science, 1913). Quote from p. 5.

(6) Reid, *Submerged*, 120.

(7) J. G. D. Clark, *The Mesolithic Settlement of Northern Europe* (Cambridge: Cambridge University Press, 1936). Clark ably summarizes the latest pollen research of the day.

(8) Bryony J. Coles, "Doggerland: A Speculative Survey," *Proceedings of the Prehistoric Society* 64 (1998): 45–81.

(9) A general summary of these developments: T. Douglas Price, "The Mesolithic of Western Europe," *Journal of World Prehistory* 1, no. 3 (1987): 225–305. See also the specialist essays in Geoff Bailey and Penny Spikins, eds., *Mesolithic Europe* (Cambridge: Cambridge University Press, 2008).

(10) The history of the Baltic Sea is summarized briefly at http://en.wikipedia .org/wiki/Yoldia_Sea.

(11) This section is based on Gaffney et al., *Europe's Lost World*, chaps. 3 and 4.

(12) The literature on Star Carr is proliferating rapidly as a result of new generations of research. J. G. D. Clark, *Excavations at Star Carr* (Cambridge: Cambridge University Press, 1954), is the classic account. Paul Mellars and Petra Dark, *Star Carr in Context* (Cambridge: McDonald Institute for Archaeological Research, 1999), is an invaluable update, even if some of its conclusions are being modified by new and as yet unpublished research.

(13) Summarized by Steven J. Mithen, "The Mesolithic Age," in *Prehistoric Europe: An Illustrated History*, ed. Barry Cunliffe (Oxford: Oxford University Press, 1994), 79–135.

Chapter 3   Euxine and Ta-Mehu

(1)   The collapse of Lake Agassiz was, of course, a much more complicated process than this, and its effect on the ocean conveyor belt is much debated. For the conveyor belt, see Wallace S. Broecker, "Chaotic Climate," *Scientific American,* November 1995, 62–68.

(2)   Graeme Barker, *The Agricultural Revolution in Prehistory* (Oxford: Oxford University Press, 2006).

(3)   Graeme Barker, *Prehistoric Farming in Europe* (Cambridge: Cambridge University Press, 1985).

(4)   A popular account of this event: William Ryan and Walter Pitman, *Noah's Flood: The New Scientific Discoveries About the Event That Changed History* (New York: Simon & Schuster, 1999). Subsequent research is questioning many details of the science. The connection with Noah's flood is pure speculation and is discounted by most scholars (including me).

(5)   Herodotus, *The Histories*, trans. Robin Waterfield (Oxford: Oxford University Press, 1998), book 2, line 5, 97.

(6)   Florence Nightingale, *Letters from Egypt: A Journey on the Nile, 1849–50* (London: Barrie and Jenkins, 1987), 37.

(7)   My accounts of ancient Egypt and the Nile are based on personal experience and on two basic sources: Barry Kemp, *Ancient Egypt: The Anatomy of a Civilization*, 2nd ed. (London: Routledge, 2006), and Toby Wilkinson, *The Rise and Fall of Ancient Egypt* (New York: Random House, 2010).

(8)   Wilkinson, *The Rise*, 24.

(9)   An excellent description: H. J. Dumont, "A Description of the Nile Basin, and a Synopsis of Its History, Ecology, Biogeography, Hydrology, and Natural Resources," in Dumont, *The Nile: Origin, Environments, Limnology and Human Use*, ed. H. J. Dumont (New York: Springer Science, 2009), 1–21. This section is

also based on Daniel Jean Stanley and Andrew G. Warne, "Nile Delta: Recent Geological Evolution and Human Impact," *Science* 260 (1993), 5108: 628–34. See also Waleed Hamza, "The Nile Delta," in Dumont, *The Nile*, 75–94.

(10) Barker, *The Agricultural Revolution*, chap. 4, has an excellent summary.

(11) A short description of the site: Josef Eiwanger, "Merimde Beni-salame," in *Encyclopedia of the Archaeology of Ancient Egypt*, ed. Kathryn A. Bard (London and New York: Routledge, 1999), 501–5.

(12) Maadi: Béatrix Mident-Reynes, "The Naqada Period," in *The Oxford History of Ancient Egypt,* ed. Ian Shaw (Oxford: Oxford University Press, 2000), 57–60.

(13) Toby Wilkinson, *Early Dynastic Egypt* (London: Routledge, 2001), has an excellent account of these developments.

Chapter 4 *"Marduk Laid a Reed on the Face of the Waters"*

(1) This quote comes from an elaborate introduction to an incantation recited in honor of Ezida, the temple of the god Nabu at Borsippa near Babylon. L. H. King, *The Seven Tablets of Creation* (San Francisco: Book Tree, 1999). Originally published in 1902, page 4 in the 1999 edition.

(2) This passage is based on Douglas J. Kennett and James P. Kennett, "Early State Formation in Southern Mesopotamia: Sea Levels, Shorelines, and Climate Change," *Journal of Island & Coastal Archaeology* 1 (2006): 67–99. See also Paul Sanlaville, "The Deltaic Complex of the Lower Mesopotamian Plain and Its Evolution Through Millennia," in *The Iraqi Marshlands: A Human and Environmental Study*, ed. Emma Nicholson and Peter Clark (London: AMAR and Politico's, 2006), 133–50.

(3) The Intertropical Convergence Zone (ITCZ), known to sailors as the doldrums, is an area around the globe near the equator where winds originating in the Northern and Southern Hemispheres come together. Normally appearing as a bank of clouds, the north-south movements of the ITCZ have short- and long-term effects on rainfall in many equatorial nations.

(4) These paragraphs draw on Kennett and Kennett, "Early State Formation," 79–85.

(5) Quoted from Gavin Young, *Return to the Marshes: Life with the Marsh Arabs of Iraq* (London: Collins, 1977), 42–43.

(6) Quote from Wilfred Thesiger, *Desert, Marsh & Mountain* (New York: HarperCollins, 1979), 106. For an assessment of the marsh ecosystem, see M. I. Evans, "The Ecosystem," in Nicholson and Clark, *Iraqi Marshlands*, 201–19.

(7) Young, *Return*, 16.

(8) The classic account of the Marsh Arabs is Wilfred Thesiger, *The Marsh Arabs* (New York: Dutton, 1964), upon which the account of these people in this chapter is based. See also S. M. Salim, *Marsh Dwellers of the Euphrates Delta* (London: Athlone Press, 1962), and Edward L. Ochsenschlager, *Iraq's Marsh Arabs in the Garden of Eden* (Philadelphia: University of Pennsylvania Museum, 2004).

(9) Samuel Kramer, *The Sumerians* (Chicago: University of Chicago Press, 1963), 77.

(10) Leonard Woolley, *Excavations at Ur* (New York: Barnes & Noble, 1954), gives a popular account of the Ur discoveries.

(11) Woolley, *Excavations*, 34.

(12) Woolley, *Excavations*, 36.

(13) C. L. Woolley, ed., *Ur Excavations*, vol. 1 (Oxford: Oxford University Press, 1927–1946), 110–11.

(14) http://www.ancienttexts.org/library/mesopotamian/gilgamesh/tab1.htm.

(15) King, *The Seven Tablets*, 4.

Chapter 5   *"Men Were Swept Away by Waves"*

(1) This section is based on: R. T. J. Cappers and D. C. M. Raemaeker, "Cereal Cultivation at Swifterbant? Neolithic Wetland Farming on the North European Plain," *Current Anthropology* 49 (2008): 385–402.

(2) Cappers and Raemaeker, "Cereal Cultivation," 388–89.

(3) An extensive Dutch literature surrounds *terpen*. This passage is based on Annet Nieuwhof, "Living in a Dynamic Landscape: Prehistoric and Proto-Historic Occupation of the Northern-Netherlands Coastal Area," in *Science for Nature Conservation and Management: The Wadden Sea Ecosystem and EU Directives*, H. Marencic et al., *Proceedings of the 12th International Scientific Wadden Sea Symposium in Wilhelmshaven, Germany, 30 March–3 April 2009* (Wilhelmshaven: Wadden Sea Ecosystem No. 26, 2010), 174–78; also on Audrey M. Lambert, *The Making of the Dutch Landscape: An Historical Geography of the Netherlands* (New York: Seminar Press, 1971), 30–31. See also Jaap Boersma, "Dwelling Mounds in the Salt Marshes—The Terpen of Friesland and Groningen," in *The Prehistory of the Netherlands*, ed. L. P. Louwe Kooijmans (Amsterdam: Amsterdam University Press, 2005), 57–560.

(4) Kooijmans, ed., *The Prehistory of the Netherlands*, 569.

(5) Gaius Plinius Secundus, known as Pliny the Elder (A.D. 23–A.D. 79), was

a Roman author, naturalist, and natural philosopher, and also an army and naval commander. His multivolume *Natural History* summarized a lifetime of observations and became a model for many subsequent works. H. Rackham, trans. *Pliny, De Historia Naturalis*, vol. 14 (Cambridge, MA: Loeb Classical Library, Harvard University Press, 1938), 1.

(6)   William Jackson Brodribb, ed. and trans., *The History of Tacitus*, book 1 (London: Macmillan, 1898), 70. Publius Cornelius Tacitus (A.D. 56–117) was a senator and historian of the Roman Empire. He wrote five works, of which *Germania* and the *Histories* are best known. Publius Vitellius was a commander under the Roman general Germanicus. Five years after the incident described by Tacitus, he successfully prosecuted the murderer of Germanicus, who died under suspicious circumstances in A.D. 19.

(7)   This section is based on Stephen Rippon, *The Transformation of Coastal Wetlands* (London: British Academy and Oxford University Press, 2000), 32–38.

(8)   Rippon, *The Transformation*, 84–90.

(9)   Rippon, *The Transformation*, 47.

(10)   *Anglo-Saxon Chronicle, Part 3: A.D. 920–1014* Accessed at http://omacl.org /Anglo/part3.html.

(11)   These paragraphs based on Lambert, *The Making*, 77–81.

(12)   The Day of Judgment featured prominently in medieval homilies, like the Blickling Homily for Easter, quoted here, preached at a monastic house near Lincoln, England, some time in the late tenth or early eleventh century. Quoted from Michael Swanton, ed., *Anglo-Saxon Prose* (Totowa, NJ: Rowman and Littlefield, 2002), 67–96.

(13)   Bruce Parker, *The Power of the Sea* (New York: Palgrave Macmillan, 2010), 62–63.

(14)   Hubert Lamb and H. H. Lamb, *Historic Storms of the North Sea, British Isles, and Northwestern Europe* (Cambridge: Cambridge University Press, 2005), is the classic source on northern European storm surges.

(15)   Lamb and Lamb, *Historic Storms*, 74.

(16)   Parker, *The Power*, 63.

(17)   Saltwater-saturated peat was dried, then burned, the ash then being soaked in more saltwater, and then filtered before being allowed to dry into small cakes. The resulting salt served to preserve herrings, as well as being used for other purposes.

(18)   Lambert, *The Making*, 120–23.

(19)   The word "polder," a tract of land protected by dikes, was first used in about 1138 in Flanders.

Chapter 6   *"The Whole Shoreline Filled"*

(1) Homer, Samuel Butler, trans., *The Iliad* (London: Longmans, Green & Co., 1898), book XIV.

(2) Strabo, *Geography*, Horace Leonard Jones, trans. (Cambridge, MA: Loeb Classical Library, Harvard University Press, 1929), book 8, 1: 31.
Recent research: John C. Kraft et al., "Harbor Areas at Ancient Troy: Sedimentology and Geomorphology Complement Homer's *Iliad*," *Geology* 31 (2003): 163–66.

(3) Scylax of Caryanda is said to have completed his *Periplus* in about 350 B.C.E. Quote in this paragraph from Elpida Hadjidaki, "Preliminary Report of Excavations in the Harbor of Phalasarna in West Crete," *American Journal of Archaeology* 92, no. 4 (1988): 467, 468.

(4) Hadjidaki, "Preliminary Report," 463–79.

(5) Strabo, *Geography*, book 9, 1.

(6) Jean-Philippe Goiranet et al., "Piraeus, the Ancient Island of Athens: Evidence from Holocene Sediments and Historical Archives," *Geology* 39 (June 2011): 531–34.

(7) Nick Marriner et al., "Holocene Morphogenesis of Alexander the Great's Isthmus at Tyre in Lebanon," *Proceedings of the National Academy of Sciences* 104, no. 22 (2007): 9218–23. Also: Nick Marriner et al., "Geoscience Rediscovers Phoenicia's Buried Harbors," *Geology* 34 (2006): 1–4.

(8) Eduard Reinhardt and Avner Raban, "The Tsunami of 13 December A.D. 115 and Destruction of Herod the Great's Harbor at Caesarea Maritima, Israel," *Geology* 34 (2006): 1061–64.

(9) David Blackman, "Ancient Harbours in the Mediterranean," *International Journal of Nautical Archaeology* 11 (1982): 79ff.

(10) A first-rate website describes recent research at Ostia and reconstructs the harbors: www.ostia-antica.org/med/med.htm#bib.

(11) Pliny the Younger, P. G. Walsh, trans., *Complete Letters* (New York: Oxford University Press, 2009), 6, 31.

(12) A magisterial history of Venice: Roger Crowley, *City of Fortune* (New York: Random House, 2011).

(13) MOSE Project: http://en.wikipedia.org/wiki/MOSE_Project.

(14) Yeduda Bock et al., "Recent Subsidence of the Venice Lagoon from Continuous GPS and Inferometric Synthetic Aperture Radar," *Geochemistry Geophysics Geosystems*, 10.1029/2011 (2012).

Chapter 7   *"The Abyss of the Depths Was Uncovered"*

(1) Gavin Kelly, *Ammianus Marcellinus: The Allusive Historian* (Cambridge: Cambridge University Press, 2008), 141.

(2) Franck Goddio et al., *Alexandria: The Submerged Royal Quarters* (London: Periplus Publishing, 1998), also *Cleopatra's Palace: In Search of a Legend* (New York: Discovery Books, 1998). See also http://franckgoddio.org.

(3) Franck Goddio et al., *Underwater Archaeology in the Canopic Region in Egypt: The Topography and Excavation of Heraklion-Thonis and East Canopus* (Oxford: Center for Maritime Archaeology, 2007).

(4) Athenaeus, C. D. Yonge, trans., *Deipnosophistae* (London: Henry Bohn, 1854), book 1, 33d. Athenaeus of Naukratis in the Egyptian delta was a Greek rhetorician and grammarian of the late second and early third centuries C.E. His *Deipnosophistae*, "dinner-table philosophers," survives in fifteen books. He himself said he was the author of a treatise on fish, but it is lost.

(5) Patrick McGovern, *Uncorking the Past* (Berkeley: University of California Press, 2009), 180–82 is the best guide.

(6) Tutankhamun's wines: See Maria Rosa-Guasch-June et al., "First Evidence for White Wine from Ancient Egypt from Tutankhamun's Tomb," *Journal of Archaeological Science* 33, no. 8 (2006): 1075–80.

(7) Barry Kemp, *Ancient Egypt: The Anatomy of a Civilization*, 2nd ed. (London: Routledge, 2006), 10.

(8) Orrin H. Pilkey and Rob Young, *The Rising Sea* (Washington, DC: Island Press, 2009), 101–116, has an excellent summary. Also Edward Maltby and Tom Barker, eds., *The Wetlands Handbook* (Oxford: Wiley-Blackwell, 2009).

(9) Alan K. Bowman and Eugene L. Rogan, *Agriculture in Egypt from Pharaonic to Modern Times* (London: British Academy, 1999), 1–32.

(10) William Willcocks, *The Nile Reservoir Dam at Assuan, and After*, 2nd ed. (London: Spon and Chamberlain, 1903).

(11) A huge literature surrounds the High Dam, much it from the years after it was built. A popular account: Tom Little, *High Dam at Aswan: The Subjugation of the Nile* (New York: John Day, 1965).

(12) Daniel Jean Stanley and Andrew G. Warne, "Nile Delta: Recent Geological Evolution and Human Impact," *Science* 260, no. 55108 (1993): 628–634. See also M. El Raey et al., "Adaptation to the Impacts of Sea Level Rise in Egypt," *Climate Research* 12 (1999): 117–28.

Chapter 8 "*The Whole Is Now One Festering Mess*"

(1) My description of Lothal is based on S. R. Rao, *Lothal and the Indus Civilization* (Bombay: Asia Publishing House, 1973). The reconstruction drawing in the frontispiece (see figure 8.1) is the basis for my scenario.

(2) The Harappan civilization is well described by Gregory Possehl, *The Indus Civilization: A Contemporary Perspective* (Walnut Creek, CA: Altamira Press, 2003). See also Jane R. McIntosh, *The Ancient Indus Valley: New Perspectives* (Santa Barbara, CA: ABC-Clio, 2008).

(3) For a general description, see Brian Fagan, *Beyond the Blue Horizon,* (New York: Bloomsbury Press, 2012), Chap. 7.

(4) W. C. Schoff, ed. and trans., *The Periplus of the Erythraean Sea: Trade and Travel in the Indian Ocean by a Merchant of the First Century* (London: Longmans, 1912). Quotes from chaps. 40 and 45, both of which are little more than a sentence.

(5) Cyclone summary: http://www.en.wikipedia.org/wiki/List_of_North _Indian_Ocean_cyclone_seasons.

(6) A harrowing description of this famine will be found in Mike Davis, *Late Victorian Holocausts* (Verso: New York, 2001), chaps. 1–3.

(7) Bruce Parker, *The Power of the Sea* (New York: Palgrave Macmillan, 2010), chap. 3, summarizes some of the material in this passage.

(8) The Satapatha Brahmana ("Brahmana of one hundred paths") is one of the prose texts describing the Vedic ritual, compiled between the eighth and sixth centuries B.C.E. The mythological sections include legends of the creation and of a great flood.

(9) An authoritative account of this storm appears in J. E. Gastrell and Henry F. Blanford, *Report on the Calcutta Cyclone of the 5th* October 1864 (Calcutta: Military Orphan Press, 1864). I also drew on Sir John Eliot, *Handbook of Cyclonic Storms in the Bay of Bengal for the Use of Sailors* (Calcutta: Meteorological Department of the Government of India, 1894), chap. 3.

(10) Quotes in these paragraphs from Eliot, *Handbook,* 143–44.

(11) Gastrell and Blanford, *Report,* 35.

(12) Gastrell and Blanford, *Report,* 38.

(13) Gastrell and Blanford, *Report,* 121.

(14) Eliot, *Handbook,* 151–62.

(15) Eliot, *Handbook,* 162.

(16) Well summarized at http://www.en.wikipedia.org/wiki/1970_Bhola_cyclone, with numerous contemporary references.

Chapter 9   The Golden Waterway

(1)   T. Healy et al., eds., *Muddy Coasts of the World: Processes, Deposits, and Function, Proceedings in Marine Science 4* (Amsterdam: Elsevier Science, 2002),

(2)   Duncan A. Vaughan et al., "The Evolving Story of Rice Evolution," *Plant Science* 174, no. 4 (2008): 394–408.

(3)   These paragraphs are based on Li Liu et al., "The Earliest Rice Domestication in China," *Antiquity* 81 (2007), 313 Accessed at http://antiquity.ac.uk/proj gall/liu1/index.html.

(4)   D. Q. Fuller et al., "Presumed Domestication? Evidence for Wild Rice Cultivation and Domestication in the Fifth Millennium BC of the Lower Yangtze Region," *Antiquity* 8, no. 1 (2007): 116–31.

(5)   Kwang-chih Chang, *The Archaeology of Ancient China*, 4th ed. (New Haven: Yale University Press, 1986), covers this material. See chap. 4, 192–212.

(6)   For the 1954 flood: http://factsanddetails.com/china.php?itemid=395&ca tid=10&subcatid=65.

(7)   Flood intervals: Fengling Yu et al., "Analysis of Historical Floods on the Yangtze River, China: Characteristics and Explanations," *Geomorphology* 113 (2009): 210–16.

(8)   Li Liu and Xingcan Chen, *State Formation in Early China* (London: Duckworth, 2003), 116–88 was the source for these paragraphs.

(9)   Marie-Claire Bergère, *Shanghai: China's Gateway to Modernity* (Palo Alto, CA: Stanford University Press, 2009).

(10) Jeffrey N. Wasserstrom, *Global Shanghai, 1850–2010* (London: Routledge, 2009).

(11) Coco Liu and ClimateWire, "Shanghai Struggles to Save Itself from the Sea," *Scientific American*, September 27, 2011. Accessed at www.scientificamerican .com/article.cfm?id=shanghai-struggles-to-save-itself-from-east-china-sea.

See also B. Wang et al., "Potential Impacts of Sea-Level Rise on the Shanghai Area," *Journal of Coastal Research*, Special Issue 14 (1998): 151–66. Zhongyuan Chen and Daniel Jean Stanley, "Sea-Level Rise on Eastern China's Yangtze Delta," *Journal of Coastal Research* 14, no. 1 (1998): 360–66.

(12) Quanlong Wei, *Land Subsidence and Water Management in Shanghai* (Delft, Netherlands: MA Thesis, Delft University of Technology, 2006). Accessed at http://www.curnet.nl/upload/documents/3BW/Publicaties/Wei.pdf.

(13) The literature is enormous. For a summary of the Three Gorges Dam and its potential consequences, see: http://www.internationalrivers.org/en/node/356 ?gclid=CKfbmsr8tK8CFYZgTAodo394HA.

Chapter 10   *"Wave in the Harbor"*

(1)   The *Nihon* is an officially commissioned history, completed in 901 C.E. Its fifty volumes cover the years 858–887. This detailed history is only in Japanese, but a general account of the tsunami of 869 appears in Kenneth Chang, "Blindsided by Ferocity Unleashed by a Fault," *New York Times*, March 21, 2011.

(2)   Bruce Parker, *The Power of the Sea* (New York: Palgrave Macmillan, 2010), 136–142.

(3)   Junko Habu, *Ancient Jomon of Japan* (Cambridge: Cambridge University Press, 2004) was the source for my description of Jomon, which is surrounded by an enormous literature.

(4)   Habu, *Ancient Jomon*, 72–76.

(5)   All of this was time consuming; witness the experience of California Indians. A California anthropologist, Walter Goldschmidt, found that 2.72 kilograms of pounded acorns processed by a Native American woman became 2.45 pounds of meal. Leaching this sample took just under four hours, about one and three quarter hours per kilogram. Jomon processing would probably have consumed as much time. See Walter Goldschmidt, "Nomlaki Ethnography," *University of California Publications in American Anthropology and Ethnology* 42, no. 4 (1951): 303–443.

(6)   Simon Kaner, "Surviving the Tsunami: Archaeological Sites of Northeastern Japan," *Current World Archaeology* 49, no. 5, 1 (2011): 25–35.

(7)   David Bressan, "Historic Tsunamis in Japan," *History of Geology*, March 17, 2011. Accessed at http://historyofgeology.fieldofscience.com/2011/03/historic-tsunamis-in-japan.html.

(8)   Nobuo Shuto, "A Century of Countermeasures Against Storm Surges and Tsunamis in Japan," *Journal of Disaster Research* 2, no. 1 (2007): 19–26.

(9)   For a discussion of sea defenses and the 2011 tsunami, see Norimitsu Onishi, "Seawalls Offered Little Protection Against Tsunami's Crushing Waves," *New York Times,* March 13, 2011.

(10) The media coverage was enormous. A useful summary that is adequate for our purposes appears at http://en.wikipedia.org/wiki/2011_Tōhoku_earthquake_and_tsunami.

(11) My account is based on Bruce Parker's superb analysis: *Power of the Sea*, chaps. 8 and 9.

Chapter 11   A Right to Subsistence

(1) A blow-by-blow description of the founding of Bangladesh and the 1970 cyclone appears in Archer K. Blood, *The Cruel Birth of Bangladesh* (Dhaka: The University Press Limited, 2002).

(2) http://www.en.wikipedia.org/wiki/1991_Bangladesh_cyclone.

(3) An excellent summary appears in *Climate Change Case Studies,* May 2009. Accessed at http://wvasiapacific.org/downloads/case-studies/Bangladesh _Cyclone_Sidr_Response.pdf.

(4) M. L. Parry et al., *IPCC Climate Change 2007: Impacts, Adaptation and Vulnerability* (Cambridge: Cambridge University Press, 2007).

(5) Golam Mahabub Sarwar, *Impacts of Sea Level Rise on the Coastal Zone of Bangladesh* (Lund, Netherlands: Lund University International Masters Program in Environmental Science, MA Thesis, 2005).

(6) IRIN Report 11.14.2011: http://www.irinnews.org/report.aspx?ReportId =75094, "BANGLADESH: Rising sea level threatens agriculture."

(7) R. Chabra, *Soil Salinity and Water Quality* (Brookfield, VT: A.A. Balkema, 1996).

(8) IRIN Report 11.14.2011.

(9) This section is based on Edmund Penning-Rowsell et al., *Migration and Global Environmental Change CS4: Population Movement in Response to Climate-Related Hazards in Bangladesh: The "Last Resort"* (London: Government Office on Science: Foresight Project on Global Environmental Migration, 2011). Accessed at http://www.icimod.org/?q=630.

Chapter 12   The Dilemma of Islands

(1) Robert McGhee, *Ancient People of the Arctic* (Vancouver, BC: University of British Columbia Press, 1996), offers an excellent general introduction to the archaeology of the far north.

(2) Owen Mason, "The Contest Between the Ipiutak, Old Bering Sea, and Birnik Polities and the Origin of Whaling During the First Millennium AD Along Bering Strait," *Journal of Anthropological Archaeology* 17, no. 3 (1998): 240–325. Quote from p. 256.

(3) http://en.wikipedia.org/wiki/Barrier_island.

(4) Owen Mason et al., *Living with the Coast of Alaska* (Durham, NC: Duke University Press, 1997). I drew on chaps. 9 and 10 in writing descriptions of Shishmaref and Naknek. For the former, see also Orrin H. Pilkey and Rob Young, *The Rising Sea* (Washington, DC: Island Press, 2009), 7–16.

(5) For early settlement of the Pacific, see Patrick Kirch, *On the Road of the Winds* (Berkeley: University of California Press, 2002). Patrick D. Nunn, *Climate, Environment and Society in the Pacific During the Last Millennium* (Amsterdam: Elsevier, 2007), offers a general description of environmental change in the Pacific, which I drew on for these sections.

(6) Tuvalu: http://www.tuvaluislands.com. See also http://www.islandvul nerability.org/tuvalu.html.

(7) Kiribati: http://www.kiribatitourism.gov.ki. For climate change, the government's official climate change site is useful: http://www.climate.gov.ki.

(8) Alliance of Small Island States: http://aosis.info.

(9) Xavier Romero-Frias, *The Maldive Islanders: A Study of the Popular Culture of an Ancient Ocean Kingdom* (Barcelona: Nova Ethnographia Indica, 2003).

(10) Quoted from Pilkey and Young, *The Rising Sea,* 20–21.

(11) Pilkey and Young, *The Rising Sea,* See also Adam Hadhazy, "The Maldives, Threatened by Drowning Due to Climate Change, Set to Go Carbon-Neutral," *Scientific American,* March 16, 2009. Accessed at http://www.scientificamerican.com /blog/post.cfm?id=maldives-drowning-carbon-neutral-by-2009-03-16.

(12) See the United Nations Development Program for the Maldives: http:// www.undp.org.mv/v2/?lid=171.

Chapter 13 *"The Crookedest River in the World"*

(1) Mark Twain, *Life on the Mississippi* (Boston: Osgood, 1883), chap. 1, 1.

(2) The literature is diffuse and specialized. See Janet Rafferty and Evan Peacock, eds., *Times River: Archaeological Syntheses from the Lower Mississippi River Valley* (Tuscaloosa: University of Alabama Press, 2008). The chapter by Carl P. Lipo and Robert C. Dunnell, "Prehistoric Settlement in the Lower Mississippi Valley," 125–67, is especially relevant. Poverty Point: Jon L. Gibson, *The Ancient Mounds of Poverty Point: Place of Rings* (Gainesville, FL: University Press of Florida, 2004).

(3) Tristram R. Kidder, "Climate Change and the Archaic to Woodland Transition (3000–2500 Cal B.P.) in the Mississippi River Basin," *American Antiquity* 71, no. 2 (2006): 195–231.

(4) This historical passage is based on John McPhee, *The Control of Nature* (New York: Farrar, Straus, and Giroux, 1990), 31ff, also 58ff. For an entertaining and often sparkling history of the Mississippi valley before the Army Corps of Engineers: Lee Sandlin, *Wicked River: The Mississippi When It Last Ran Wild* (New York: Vintage Books, 2008).

(5) McPhee, *The Control of Nature*, 57.

(6) McPhee, *The Control of Nature*, 58.

(7) A. Baldwin Wood (1879–1956) as an inventor and engineer. His highly efficient, low-maintenance pumps drained, and still drain, much of New Orleans and have also been widely used elsewhere, including in the drainage of the Zuiderzee in the Netherlands.

(8) Discussion of Morgan City based on McPhee, *The Control of Nature*, 78ff.

(9) Inevitably, Hurricane Katrina has generated an enormous popular and academic literature. A fascinating film review essay is worth reading to get the flavor of the controversies. Nicholas Lemann, "The New New Orleans," *New York Review of Books*, March 24, 2011. A recent very solid account: James Patterson Smith, *Hurricane Katrina: The Mississippi Story* (Jackson, MS: University Press of Mississippi, 2012).

(10) Migration: Susan L. Cutter et al., "The Katrina Exodus: Internal Displacements and Unequal Outcomes" (London: Government Science Office: Foresight Migration and Global Environment Project, 2011), Case Study 1.

(11) Lehmann, "The New New Orleans," 47.

Chapter 14 *"Here the Tide Is Ruled, by the Wind, the Moon and Us"*

(1) Guntram Riecken, "Die Flutkatastrophe am 11. Oktober 1634—Ursachen, Schäden und Auswirkungen auf die Küstengestalt Nordfrieslands," in *Flutkatastrophe 1634: Natur, Geschichte, Dichtung*, 2nd ed., ed. Boy Hinrichs, Albert Panten, and Guntram Riecken (Neumünster: Wachholtz, 1991), 11–64. Quote from p. 35.

(2) Quotes in this paragraph from Riecken, "Die Flutkatastrophe," 11–12.

(3) Quoted from http://www.en.wikipedia.org/wiki/Burchardi_flood, where a general account of the disaster may be found, with primary references.

(4) This passage is based on Audrey M. Lambert, *The Making of the Dutch Landscape: An Historical Geography of the Netherlands* (New York, Academic Press), 94–102.

(5) Based on Lambert, *The Making*, 210–12.

(6) William the Silent, Prince of Orange (1533–1584), was a wealthy nobleman who rebelled against the Spanish. The Dutch revolt triggered the Eighty Years' War that ended in independence for the Republic of the Seven United Provinces in 1581, which ultimately became the Netherlands. He was assassinated in 1584.

(7) Quoted from Lambert, *The Making*, 213.

(8) Lambert, *The Making*, 213–15.

(9) Lambert, *The Making*, 215–17.

(10) Lambert, *The Making,* 218, 220.

(11) This passage is based on Lambert, *The Making,* 239–41.

(12) A. G. Maris, M. Dendermonde, and H. A. M. C. Dibbits, *The Dutch and Their Dikes* (Amsterdam: De Bezige Bij, 1956), 66.

(13) Lambert, *The Making,* 266–69.

(14) www.deltawerken.com/Zuider-Zee-flood-(1916)/306.html.

(15) For the British disaster, see Hilda Grieve, *The Great Tide: The Story of the 1953 Flood Disaster in Essex* (Colchester: Essex County Council, 1959). A summary of the Netherlands story appears at www.deltawerken.com/89.

(16) This account and quote based on personal observation, www.deltawerken.com/23 and http://en.wikipedia.org/wiki/Oosterscheldekering.

(17) Account based on personal observation.

Epilogue

(1) Digital databases: Even at this early stage in research, the digital information in the databases is sufficient to identify elevations of individual land parcels the size of a small house lot, so the data is far more accurate than any earlier assessments. Cynthia Rosenzweig et al., *Climate Change and Cities* (Cambridge: Cambridge University Press, 2011).

(2) J. L. Weiss et al., "Implications of Recent Sea Level Rise Science for Low-Elevation Areas in Coastal Cities of the Conterminous U.S.A.," *Climate Change* 105 (2011): 635–45.

(3) Dan Cayan et al., *Climate Change and Sea Level Rise Scenarios for California Vulnerability and Adaptation Assessment* (Sacramento: California Natural Resources Agency, 2012).

(4) Weiss et al., "Implications," 635–45.

(5) http://www.huffingtonpost.com/2012/10/30/hurricane-sandy-cuomo-bloomberg-climate-change_n_2043982.html

(6) Orrin H. Pilkey and Rob Young, *The Rising Sea* (Washington, DC: Island Press, 2009), 4.

(7) The definitive study is Foresight's *Migration and Global Environmental Change* (London: Government Office on Science, Foresight Project on Global Environmental Migration, 2011).

(8) Mike Davis, *Late Victorian Holocausts* (New York: Verso, 2001), chap. 11, describes Chinese famines and their consequences. Brian Fagan, *Floods, Famines, and Emperors* (New York: Basic Books, 2009), chap. 6, summarizes what is known of ancient Egyptian famines.

Index

A Note on the Author

Brian Fagan is emeritus professor of anthropology at the University of California–Santa Barbara. Born in England, he did fieldwork in Africa, and has written about North American and world archaeology and many other topics. His books on the interaction of climate and human society have established him as the leading authority on the subject; he lectures frequently around the world. He is the editor of *The Oxford Companion to Archaeology* and the author of *Beyond the Blue Horizon, Elixir, Cro-Magnon, The Great Warming, Fish on Friday, The Little Ice Age,* and *The Long Summer,* among many other titles.